Force Presentation in U.S. Air Force History and Airpower Narratives

Alan J. Vick

Prepared for the United States Air Force

For more information on this publication, visit www.rand.org/t/RR2363

Library of Congress Cataloging-in-Publication Data is available for this publication.
ISBN: 978-1-9774-0006-2

Published by the RAND Corporation, Santa Monica, Calif.

Preface

Each branch of the U.S. armed services has created organizational constructs, processes, and strategic narratives to accomplish six key functions: (1) sizing forces, (2) deploying forces, (3) employing forces, (4) sustaining operational effects, (5) managing force rotations, and (6) articulating service purpose. In the U.S. Air Force (USAF), these functions are typically grouped together under what is known as *force presentation.* Although force presentation is most formally associated with the fifth function (managing force rotations), USAF leaders and personnel often use it in reference to all six functions.

The director of Strategy, Concepts and Assessments, Headquarters, USAF, commissioned a fiscal year 2016 RAND Project AIR FORCE study to help the USAF improve its force presentation model and associated narrative. This report, which is one of the products of that study, seeks to answer three questions: (1) What is the current USAF force presentation construct, and how did it evolve historically? (2) What USAF force presentation constructs are most accessible to external audiences? (3) How might force presentation be integrated into the USAF strategic narrative?

The research described in this report was conducted within the Strategy and Doctrine Program of RAND Project AIR FORCE.

RAND Project AIR FORCE

RAND Project AIR FORCE (PAF), a division of the RAND Corporation, is the U.S. Air Force's federally funded research and development center for studies and analyses. PAF provides the Air Force with independent analyses of policy alternatives affecting the development, employment, combat readiness, and support of current and future air, space, and cyber forces. Research is conducted in four programs: Force Modernization and Employment; Manpower, Personnel, and Training; Resource Management; and Strategy and Doctrine. The research reported here was prepared under contract FA7014-16-D-1000.

Additional information about PAF is available on our website: http://www.rand.org/paf

This report documents work originally shared with the U.S. Air Force on October 5, 2016. The draft report, issued on August 31, 2017, was reviewed by formal peer reviewers and U.S. Air Force subject-matter experts.

Contents

Figures

Tables

Summary

Every branch of the armed forces has at least one construct for force presentation, though only the U.S. Air Force (USAF) uses this particular terminology and there is no authoritative definition of the concept in USAF or joint doctrine. Distilled to its essence, force presentation is the preferred organizational construct through which a service offers its capabilities to the combatant commander (CCDR). Force presentation constructs revolve around the core combat units of each service.[1] In 2018 these are the Marine Air-Ground Task Force (MAGTF),[2] the U.S. Navy's Carrier Strike Group (CSG), the U.S. Army's Brigade Combat Team (BCT), and the USAF's Air Expeditionary Task Force (AETF).

These constructs reflect, to varying degrees, each service's understanding of its unique contribution to national defense, its tactical doctrine, its historical experience, its readiness model, and the organizational metric that it believes makes the most compelling case for its total force structure. Force presentation constructs are increasingly salient to service leaders because they are often viewed as a way to constrain CCDR demands for rotational forces and personnel, thereby protecting service readiness and retention. USAF leaders, in particular, are concerned that their force presentation construct is lacking in this regard.

This report presents historical analysis and recommendations to inform USAF deliberations regarding future force presentation constructs. In particular, it offers empirically grounded answers to three questions prerequisite to any revision of USAF force presentation concepts:
(1) What is the current USAF force presentation construct, and how did it evolve historically?
(2) What USAF force presentation constructs are most accessible to external audiences?
(3) How might force presentation be integrated into the USAF strategic narrative?

Research findings and recommendations are presented below.

Findings

Force Presentation Plays Distinct and Varied Roles in the Four Services

This analysis has found that each of the services uses more than one construct to accomplish six key functions: (1) sizing forces, (2) deploying forces, (3) employing forces, (4) sustaining operational effects, (5) managing force rotations, and (6) articulating service purpose. For

---

[1] For example, in the USAF, force presentation emphasizes the CAF and at times underrepresents or fails to represent other service capabilities such as mobility, space, cyber, communications, and combat support.

[2] Although the MAGTF is the scalable framework that the USMC uses to build, command, and control task forces, the standard manifestation is the MEU/ARG. Both are force presentation constructs. See Christopher P. Cavas, "Top Marine: No Need to Change Deploying Groups," *Defense News*, August 10, 2016.

example, the Army uses the BCT construct to deploy and employ forces, the division and corps to sustain effects, the division to size the total force, and the Army Force Generation (ARFORGEN) process to manage force rotations.

Force presentation is arguably most central to U.S. Marine Corps (USMC) and U.S. Navy (USN) institutional identities and cultures because they articulate their service purpose primarily as providing forward presence and crisis response through rotational deployments of the Carrier Strike Group/Expeditionary Strike Group (CSG/ESG) and Marine Expeditionary Unit/Amphibious Readiness Group (MEU/ARG), respectively. In contrast, neither the Army nor the USAF use force presentation constructs to articulate their service purpose.

The Squadron Has Been the Most Common USAF Force Presentation Construct for the Combat Air Forces

A comparison of the squadron with the group, wing, or various AEF-related constructs over the history of the USAF finds the permanent Combat Air Forces (CAF) squadron to be the most frequently used across the six functions assessed in this report. The squadron has been the preferred deployment construct for all major conflicts, was paired with the group or wing for force employment, and was the most common force-sizing metric for the fighter and bomber force. The squadron appeared as the force presentation construct either alone or in tandem with the group or wing in 62 percent of the cases considered in this report. Although the squadron is central to USAF culture and operations and possesses a surface familiarity to outside audiences, it is probably not the best vehicle for public outreach. As will be discussed below, for most people the squadron and other tactical echelons are abstractions that by definition have less visceral power than physical objects like aircraft, ships, and tanks.

Force Presentation Has Been Largely Ignored by Airpower Theorists and Historians

Force presentation concepts are largely absent from contemporary USAF planning documents, as well as the writings of prominent airpower theorists. Force presentation is also rarely discussed in operational histories of the USAF at war. Although organization is of practical concern to all military organizations and leaders, the most prominent airpower theorists (William "Billy" Mitchell, Alexander P. de Seversky, John Warden, and David Deptula) have conceptualized and articulated the contribution of airpower using an effects-to-missions-to-targets-to-forces framework. This framework typically ends by highlighting specific airframes (e.g., the B-2), classes of aircraft (e.g., stealth bombers), and weapons types (e.g., the Joint Air-to-Surface Standoff Missile, or JASSM) among the required forces. This is helpful from a programmatic perspective, but is incomplete as a narrative because it leaves out both force presentation and basing (a critical enabler of global reach). Histories of the USAF at war also usually ignore organization except for descriptive purposes. Rather, they seek to tell a story that largely follows the effects-to-missions-to-targets-to-forces framework. They describe the air campaign objectives, planning process, and key participants; identify the critical targets; and tell the story of air combat, typically centered around particular individuals and platforms.

USAF Aircraft Are the Most Visible and Accessible Manifestation of Airpower

The modern USAF is an air, space, and cyber force, but news accounts of USAF activities typically focus on the deployment or movement of aircraft. An August 14, 2016, *National Interest* article titled "B-1, B-2, and B-52 Bombers All Descend on Guam in a Massive Show of Force" is representative of news coverage of USAF deployments. Airmen, understandably, want to tell a broader story, but there is no getting around the visceral power of iconic aircraft. As noted in Chapter Four, Google searches for aircraft like the F-22 greatly outnumber searches for the Air Expeditionary Wing (AEW), the BCT, the CSG, or the MEU. Specialized audiences will care that the USAF presents such forces as the Air Expeditionary Squadron (AES), the AETF, or the AEW, but the broader public has little use for such hard-to-visualize abstractions.

Recommendations

This research leads to three recommendations for USAF leaders:

- **Set realistic goals for any modifications to force presentation constructs.** This research showed that no single construct can address all six functions. Modifications should focus on better articulating USAF capabilities regarding particular functions such as "size forces" in a manner consistent with USAF history, culture, doctrine, and practices. Each service has developed force presentation concepts appropriate for its unique circumstances; none is easily transferable to other services. More broadly, a construct or set of constructs created by any one of the military branches is a relatively weak reed in the Global Force Management (GFM) system. Requests from CCDRs for forces will always have more weight in this process than efforts by the services to constrain demands. If USAF and other service leaders believe that CCDR demands for the CAF are unsustainable, they would be better served acting in concert to reform the GFM process so that it better highlights the readiness and other costs associated with current operational tempo.
- **Incorporate force presentation and basing in the USAF strategic narrative.** The most public American airpower narratives from the interwar years to today (i.e., those of Mitchell, de Seversky, Warden, and Deptula) have followed a framework that begins with desired effects, derives missions and targets from those goals, and ends by identifying the forces needed to destroy those targets and accomplish the specified missions. This framework is logical but does not tell the whole story, leaving out both force presentation and basing. With regard to basing, the USAF is able to generate forces with such agility for a surprising reason: due to its global network of fixed facilities. From these locations the USAF integrates effects across domains via Air Operations Centers, conducts cybermissions, launches and controls space systems, and generates aircraft sorties. Thus a more complete airpower narrative would use the following logic: desired effects to missions to targets to forces to force presentation to bases (see Figure 4.2).

- **In public outreach, emphasize agility and scalability rather than organizational abstractions.** The best USAF force presentation public narrative is a simple one that involves no organizational references at all; the USAF is the most agile and scalable military force, able to respond rapidly and globally in packages as small as a single aircraft or as large as many hundreds of platforms supported by thousands of personnel. It does this through various organizational structures and processes, including the AEF and the widespread use of Unit Type Codes (UTCs). But these "how" elements distract from the "what" and "why" narratives of widest interest and appeal. Agility and scalability are easily communicated concepts if connected to well-known physical objects like aircraft or satellites. Distinctive aircraft like the F-22 or B-2 are especially powerful icons and useful to illustrate agility in action. Organizational abstractions, no matter how important for internal processes, are no match for a clear, concise, and concrete narrative centered on USAF personnel deploying and employing advanced technologies on behalf of the nation.

Acknowledgments

The author thanks A5S for sponsoring this study. Its members graciously met with the project team on a regular basis and provided much appreciated guidance, feedback, and encouragement. Mr. Pat Harding and Mr. Joseph Pak in HAF/A5SM offered constructive feedback and assistance over the course of this study. Colonel Robert Levin, HAF/A5SM, provided helpful comments on the draft report. Thanks also to Mr. Sam Szvetecz in HAF/A5SW for proposing the study, for sharing his insights on the topic, and for his comments on the study's final briefing.

Project members Cristina Garafola, Phillip Johnson, Stacie Pettyjohn, and Meagan Smith all provided helpful comments on earlier versions of this work.

Col Michael Pietrucha, USAF, offered a valuable operator's perspective on the impact of split operations on flying squadrons (both deployed and home station elements).

Thanks to Dr. Daniel Haulman at the Air Force Historical Research Agency, Lt Gen Allen Peck (USAF, retired), and Lt Gen Steven Kwast at Air University for sharing their insights on the history of USAF organizations, the AEF in action and future directions for airpower. Staff from ARFORPAC; Marine Corps Forces, Pacific; Pacific Air Forces; and U.S. Pacific Command, met with the project team to discuss the role of combatant commands and components in the GFM process.

Many RAND colleagues shared their deep expertise on a wide range of topics. Thanks especially to Lily Ablon, Katharina Best, Irv Blickstein, Bryan Boling, Mike Decker, Paul Dreyer, Derek Eaton, Gian Gentile, John Gordon, Lisa Harrington, David Johnson, Darrell Jones, Jennifer Kavanagh, Sherill Lingel, Bradley Martin, Pat Mills, Igor Milolic-Torreira, David Ochmanek, Lara Schmidt, Don Snyder, and Paula Thornhill.

Gian Gentile, Phillip Meilinger, and Tom McNaugher provided expert, insightful, and constructive reviews of the draft report.

Finally, thanks to Rosa Meza and Donna White for preparing the manuscript and to production editor Kimbria McCarty for her skillful management of the editing and publications process.

Abbreviations

AAS	Army Air Service
ADC	Air Defense Command
AEF	Air Expeditionary Force
AEG	Air Expeditionary Group
AES	Air Expeditionary Squadron
AETF	Air Expeditionary Task Force
AEW	Air Expeditionary Wing
AFCON	Headquarters, Air Force Controlled
AFHRA	Air Force Historical Research Agency
AFI	Air Force Instruction
ARFORGEN	Army Force Generation (model)
ARG	Amphibious Readiness Group
ATO	Air Tasking Order
BCT	Brigade Combat Team
CAF	Combat Air Forces
CASF	Composite Air Strike Force
CCDR	combatant commander
CCTS	Combat Crew Training Squadron
CENTCOM	U.S. Central Command
CLF	Combat Logistics Force
CONUS	continental United States
CSG	Carrier Strike Group
CVN	aircraft carrier, fixed-wing, nuclear-powered
CVW	Carrier Air Wing
DOD	Department of Defense
EAF	Expeditionary Air Force

ESG	Expeditionary Strike Group
FEAF	Far East Air Force
GFM	Global Force Management
GWAPS	*Gulf War Air Power Survey*
ISR	intelligence, surveillance, and reconnaissance
JFACC	Joint Force air component commander
JOPES	Joint Operational Planning and Execution System
MAGTF	Marine Air Air-Ground Task Force
MATS	Military Air Transport Service
MEB	Marine Expeditionary Brigade
MEF	Marine Expeditionary Force
MEU	Marine Expeditionary Unit
NFZ	no-fly zone
O-FRP	Optimized Fleet Response Plan
ONW	Operation Northern Watch
OSD	Office of the Secretary of Defense
OSW	Operation Southern Watch
PAF	RAND Project AIR FORCE
SAC	Strategic Air Command
TAC	Tactical Air Command
TFW	Tactical Fighter Wing
USAAF	U.S. Army Air Forces
USAF	U.S. Air Force
USMC	U.S. Marine Corps
USN	U.S. Navy
UTC	Unit Type Code

1. Introduction

Every branch of the armed forces has at least one construct for force presentation, though only the U.S. Air Force (USAF) uses this particular terminology and there is no authoritative definition of the concept in USAF or joint doctrine.[1] Distilled to its essence, force presentation is the preferred organizational construct through which a service offers its combat capabilities to the combatant commander (CCDR). Force presentation constructs revolve around the core combat units of each service.[2] In 2016, these were the Marine Air-Ground Task Force (MAGTF),[3] the U.S. Navy's Carrier Strike Group (CSG), the U.S. Army's Brigade Combat Team (BCT) and the USAF's Air Expeditionary Task Force (AETF).

These constructs reflect, to varying degrees, each service's understanding of its unique contribution to national defense, its tactical doctrine, its historical experience, its readiness model, and the organizational metric that it believes makes the most compelling case for its total force structure. Force presentation constructs are increasingly salient to service leaders because they are often viewed as a way to constrain CCDR demands for rotational forces and personnel, thereby protecting service readiness and retention. USAF leaders, in particular, are concerned that their force presentation construct is lacking in this regard.

Although the current focus on the role of force presentation in managing frequent, relatively short deployments of forces is understandable, it represents an overly narrow understanding of the concept. It is also a relatively new focus. Although the United States deployed naval forces overseas as early as 1798 and based both naval and ground forces abroad following the acquisition of new U.S. territory after the Spanish-American War of 1898, it was not until after the Cold War ended that the Department of Defense (DOD) embraced a strategy

[1] The term *force presentation* does not appear in the DOD dictionary, where it would if it had standing in joint doctrine. USAF publications make reference to it, but the meaning seems to be inferred or assumed; see, e.g., U.S. Air Force, *Air Force Instruction 10-401, Air Force Operations Planning and Execution*, Washington, D.C.: Department of the Air Force, December 7, 2006; and U.S. Air Force, *Annex 3-30 Command and Control*, Maxwell Air Force Base, Ala.: Curtis E. LeMay Center for Doctrine Development and Education, 2014e. The closest to a definition is found in the USAF *Strategic Master Plan*: "Force presentation refers to how the Air Force provides forces and support (equipment and resources) to meet global CCDR requirements. The Air Force supports global CCDR requirements through a combination of assigned, attached (rotational), and mobility forces that may be forward deployed, transient, or operating from home station." U.S. Air Force, *Strategic Master Plan: Strategic Posture Annex*, Washington, D.C.: Department of the Air Force, May 2015. See also Joint Chiefs of Staff, *Department of Defense Dictionary of Military and Associated Terms*, Washington, D.C.: Joint Chiefs of Staff, JP 1-02, November 8, 2010 (amended through February 2016).

[2] For example, in the USAF, force presentation emphasizes the CAF and at times underrepresents or fails to represent other service capabilities such as mobility, space, cyber, communications, and combat support.

[3] Although the MAGTF is the scalable framework that the USMC uses to build, command, and control task forces, the standard manifestation is the MEU/ARG. Both are force presentation constructs. See Christopher P. Cavas, "Top Marine: No Need to Change Deploying Groups," *Defense News*, August 10, 2016.

dependent on frequent rotations of forces for deployments lasting three to sixth months.[4] From 1898 until the end of the Cold War, U.S. forces were based abroad in permanent garrisons. Personnel were rotated individually with tours typically lasting from one year (for unaccompanied tours) to three years (for accompanied tours). Thus, managing force rotations abroad was not a central concern for U.S. Marine Corps (USMC) and U.S. Navy (USN) leaders until the Cold War (when those services embraced forward presence strategies), nor was it a priority for the USAF until extended stability operations became the norm in the 1990s. Finally, the U.S. Army created the Army Force Generation (ARFORGEN) process to manage brigade rotations only after the demands of operations in Afghanistan and Iraq threatened to break the force.[5]

As will be discussed in more detail later in this introduction, this study identified not one but six service functions where force presentation is a consideration. Indeed, none of the services has a single construct but rather between five and six, depending on the functional need. These functions are not equally important to every service, but an understanding of how force presentation is manifested in each function and across the services is essential for service leaders who wish to better align their construct(s) with national strategy, ongoing operations, and future challenges. With respect to managing rotations, there are important differences among the services. It is appropriate for the USMC and the USN to emphasize this one aspect of force presentation because those services articulate their value to the nation primarily as a function of their rotational forward presence and crisis response capabilities. In contrast, force presentation constructs are largely absent from the Army and Air Force narratives.

A Cross-Service Framework for Assessing Force Presentation Constructs

Every service needs organization constructs, concepts, and processes to accomplish six key functions: (1) sizing forces, (2) deploying forces, (3) employing forces, (4) sustaining operational effects, (5) managing force rotations, and (6) articulating service purpose. Although these functions are largely self-explanatory, a few words are in order regarding *force rotations*; this report uses that term to describe the process by which a service manages an ongoing demand for temporary deployments of force elements such as squadrons, brigades, Marine Expeditionary Units (MEUs), or carriers, typically on the order of six months or less. Such demands were largely unheard of for the Army and USAF during the Cold War. Today they are a source of great stress for all the services. Thus, for the purposes of this report, the

---

[4] For a history of U.S. military forward deployments from the founding of the nation see Stacie L. Pettyjohn, *U.S. Global Defense Posture, 1783–2011*, Santa Monica, Calif.: RAND Corporation, MG-1244-AF, 2012.

[5] The Vietnam War presented its own rotational challenges due to the individual replacement policy in which military personnel, with a few exceptions, rotated in and out of units on one-year tours. Individual divisions and wings did not, however, rotate subordinate elements on a regular basis.

various mechanisms the services used over the course of American history to maintain permanent forces abroad, rotate individual replacements during wartime, or deploy major formations to large wars (e.g., World War II) are not force rotations.

As Table 1.1 illustrates, no service has a single construct that works for all six functions.

Table 1.1. Cross-Service Force Presentation Framework, 2016

	Size Forces (2014 QDR)	*Deploy* Forces	*Employ* Forces	*Sustain* Operational Effects	*Manage* Force Rotations	*Articulate* Service Purpose
USN	CVNs, CVWs, combatant vessels	CSG/ESG	CSG/ESG	Combat Logistics Force	O-FRP	CSG//ESG offers presence & crisis response
USMC	MEFs, Divisions, Air Wings	MEU	MEU	MAGTF & ARG	R-Plus or D-Minus	MEU/ARG offers presence & crisis response
Army	Divisions	BCT	BCT	Division & Corps	ARFORGEN	FP plays no role in service narrative
USAF	Squadrons	UTC, AES, Squadron	Squadron, wing, AES, AEG, AEW, AETF	AEG, AEW or Wing	AEF	FP plays no role in service narrative

NOTE: The Marine Expeditionary Force (MEF) and Marine Expeditionary Force (MEU) are specific types of MAGTFs; R-Plus and D-Minus are used in the USMC to distinguish between two force management approaches.

As noted above, the Carrier Strike Group (CSG) and Expeditionary Strike Group (ESG) are the most important constructs for the Navy and central to its service narrative but not used for every function. For example, the Optimized Fleet Response Plan (O-FRP) is the management tool used to manage rotations, and the Combat Logistics Force (CLF) sustains the CSG; the Navy is sized not by CSGs, but by carriers, carrier air wings, and combatant vessels. For the USMC, the MAGTF provides a scalable construct for deployments from the battalion to the division level. The Marine Expeditionary Unit/Amphibious Readiness Group (MEU/ARG) is the most common version of the MAGTF. The Army shifted from a division-centric to a brigade-centric force over the last 15 years or so, but divisions and corps remain critical to sustaining operations and could regain their dominance in a larger conflict. The ARFORGEN process is the mechanism by which the Army manages rotations abroad.

Similar to the other services, the USAF uses a mix of constructs. It deploys forces using the Unit Type Code (UTC) and squadron constructs. Squadrons and wings are the key organizations for force employment in the Combat Air Forces (CAF), though they are most likely to be employed as expeditionary squadrons operating as part of an AETF or Air Expeditionary Wing (AEW). Since squadrons are not self-sustaining organizations, enduring operations require a wing-level organization for maintenance, supply, medical, security, and other functions. For ongoing operations like Operation Inherent Resolve, sustaining operation effects is provided by an AETF or AEW. In contrast, as of the 2014 *Quadrennial Defense Review* (*QDR*), the USAF used the squadron to size the total force (although it has used groups and wings in the past for the same purpose). The USAF currently manages force rotations through the Air Expeditionary Force (AEF) process. The AEF determines which units and personnel are available for deployment on a recurring 18-month schedule. Finally, although force presentation is central to the USMC and USN strategic narratives, it plays no role in USAF narratives. (See Chapter Four for more detail.)

The Purpose of This Document

This report offers historical analysis and recommendations to inform USAF deliberations regarding future force presentation constructs. In particular it offers empirically grounded answers to three questions prerequisite to any revision of USAF force presentation concepts: (1) What is the current USAF force presentation construct, and how did it evolve historically? (2) What USAF force presentation constructs are most accessible to external audiences? (3) How might force presentation be integrated into the USAF strategic narrative?

Organization

The report is loosely organized around the three questions listed above. Chapter Two explores force presentation in USAF history from its origins in the U.S. Army Signal Corps to the large role the U.S. Army Air Forces played in World War II. Chapter Three continues this historical examination, focusing on USAF force presentation from 1946 to 2016. Chapter Four answers the last two questions: What USAF force presentation constructs are most accessible to external audiences? How might force presentation be integrated into the USAF strategic narrative? Chapter Five presents conclusions and recommendations.

2. USAF Force Presentation Constructs from 1913 to 1945

Overview

This chapter explores the evolution of USAF force presentation constructs from the early days of the Aeronautical Division of the U.S. Army Signal Corps to the end of World War II. Each historical section discusses how the USAF organized to deploy and employ forces, sustain operational effects, and size forces.[1]

One theme that runs through this chapter is the close relationship among squadrons, groups, and wings in creating and sustaining combat effects. USAF veterans and serving personnel know this through personal experience but, surprisingly, there is no single scholarly work that documents the relative responsibilities and relationships of these organizations over time.[2] USAF historians and airpower scholars have paid relatively little attention to these organizational and process issues, though there are notable exceptions, such as the *Gulf War Air Power Survey* and the work of Maurer Maurer and Fred Shiner. There is no single work that traces the evolution of organizational changes and their implications for force employment and airpower narratives. This study therefore had to rely on a collection of historical sources (official histories, academic work, autobiographies) and contemporary USAF doctrinal publications to document these relationships. It also is notable that that there do not appear to be any Air Force Instructions (AFIs) or doctrinal publications that explicitly discuss the relative responsibilities of squadrons and wings in, for example, mission planning and execution.[3]

Throughout its history the USAF has presented forces primarily as squadrons, groups, wings, and task forces.[4] Groups and wings may be permanent or provisional (temporary)

---

[1] The observant reader will note that the use of force presentation to articulate service purpose is not included. Force presentation constructs have not been used to tell the USAF story and are absent from airpower narratives. See Chapter Four for details.

[2] At least the author was unable to find anything along those lines. The author also asked a former USAF historian and the current chief of the organizational history branch at the Air Force Historical Research Agency (AFHRA) if they knew of a comprehensive historical treatment of tactical organizations or force presentation and they both answered no. (AFHRA did provide a number of helpful documents that are cited in this report.)

[3] Although it is not conclusive proof that there are no such publications, it is of interest that a 2003 Air University Press book offering guidance to future squadron commanders does not cite a single USAF AFI or doctrinal publication on any aspect of squadron leadership. See Jeffry F. Smith, *Commanding an Air Force Squadron in the Twenty-First Century: A Practical Guide of Tips and Techniques for Today's Squadron Commander*, Maxwell Air Force Base, Ala.: Air University Press, 2003.

[4] The USAF also uses UTCs from the Joint Operation Planning and Execution System. The UTC is the "basic building block used in force planning and the deployment of AETFs. . . . A UTC depicts a force capability with personnel and/or equipment requirements." This enables deployments below the squadron level, down to individuals if necessary, greatly increasingly USAF flexibility in meeting the near constant demands to support overseas operations since the end of the Cold War. See U.S. Air Force, *Air Force Instruction 10-401, Air Force Operations Planning and Execution*, Washington, D.C.: Department of the Air Force, December 7, 2006, p. 154.

organizations. Task forces are, in theory, temporary organizations, though some joint task forces have become semipermanent.[5] Squadrons are, in contrast, enduring and, with rare exceptions, permanent organizations; they are the one constant in USAF force presentation history. To be sure, squadrons rarely operate as purely independent organizations. Rather, they almost always reside within a larger group or wing structure, organizations that fill vital roles both in peace and war. Nevertheless, it is the squadron that has played an outsize role in USAF culture from its beginning, which is where we will start our survey of force presentation in USAF history.

Origins

The roots of contemporary USAF concepts for force presentation are found in U.S. Army efforts to create an aviation organization, initially as the Aeronautical Division of the Army Signal Corps, founded in 1907.[6] In 1911, as the Army began to build a small aviation force, 1LT Benjamin D. Foulois, who later became Chief of the Air Corps,[7] organized three Wright Flyer aircraft and three pilots into an "aviation company" at Fort Sam Houston, Texas.[8] Two years later, on March 5, 1913, the Army established the 1st Aero Squadron (provisional) at Texas City, Texas. The squadron was composed of nine aircraft, nine officers (pilots), and 51 enlisted personnel organized into two companies.[9] Just five months after the creation of the first aviation squadron in American history, the Army also used the squadron as its first force-sizing metric for aviation. During August 1913 hearings on the Act to Increase the Efficiency in the Aviation Service, the House Committee on Military Affairs heard testimony from both

---

[5] The Joint Interagency Task Force South is an example of a semipermanent joint task force, first established in 1994. See Joint Interagency Task Force South, homepage.

[6] This structure lasted from 1907 to 1914, when the Aeronautical Division was changed to the Aviation Section, both in the Signal Corps. The first big change was in 1918, when the Army Air Service was created as a separate entity outside the Signal Corps. In 1926 the Army Air Service was replaced by the Army Air Corps. In 1941, the Army Air Forces became responsible for all Army aviation (with the Air Corps continuing as a subordinate element until 1947). Finally, on September 18, 1947, the U.S. Air Force was created as an independent military branch. See U.S. Air Force, *Fact Sheet: 1907–1947—The Lineage of the U.S. Air Force*, Washington, D.C.: U.S. Air Force Historical Support Division, 2011b.

[7] For more on Foulois's career, see Benjamin D. Foulois and C. V. Clines, *From the Wright Brothers to the Astronauts: The Memoirs of Major General Benjamin D. Foulois*, New York: McGraw-Hill, 1968; and John F. Shiner, "Benjamin D. Foulois: In the Beginning," in John L. Frisbee, *Makers of the United States Air Force*, Washington, D.C.: Pergamon-Brassey's, 1989.

[8] Roger G. Miller, *A Preliminary to War: The 1st Aero Squadron and the Mexican Punitive Expedition of 1916*, Washington, D.C.: Air Force History and Museums Program, 2003, p. 4.

[9] Miller, 2003, p. 5; Daniel Haulman, *Lineage and Honors History of the 1st Reconnaissance Squadron (ACC)*, Maxwell Air Force Base, Ala.: Air Force History Research Agency, November 2013; Bernard C. Nalty, ed., *Winged Shield, Winged Sword: A History of the United States Air Force:* Vol. I, *1907–1950*, Washington, D.C.: Air Force History and Museums Program, 1997, p. 28; Juliette A. Hennessy, *The United States Army Air Arm: April 1861 to April 1917*, Washington, D.C.: Office of Air Force History, 1985, p. 76.

senior and junior aviation officers, including CPT William Mitchell and 1LT Henry H. Arnold, among others. BG George P. Scriven, the chief signal officer, presented a plan for the growth of Army aviation. The plan would align a squadron with each of the Army's six divisional commands. Because the divisions were understrength in 1913, the plan envisioned only four squadrons initially and became known as the Four Squadron Plan.[10]

In 1914, Foulois, now a captain, took command of the 1st Aero Squadron, replacing the company-level sub-elements with "sections." Air Force historian Roger Miller describes the new squadron organization:

> In May, he abandoned the company organization and established a more flexible section organization, which included headquarters, supply, engineer, and transportation sections and eight airplane sections, one for each airplane. Under the new organization, two officers—a pilot and an assistant pilot—were assigned to each airplane. Each pilot took responsibility for care, repair, and maintenance of his airplane and the training and discipline of the crew.[11]

Sometime between 1914 and the U.S. entry into World War I, the section structure was replaced with flights as sub-elements.[12] The squadron and flight have remained as the fundamental building blocks of the USAF since. When flights require further subdivisions, they are broken into *sections*, with *elements* underneath sections.[13]

The squadron was also the first aviation force presentation construct for combat. Following the March 1916 attack on Columbus, New Mexico, by Mexican revolutionaries under the command of Pancho Villa, President Woodrow Wilson ordered BG John Pershing to lead the Punitive Expedition into Mexico to capture Villa. The 1st Aero Squadron, under the command of Foulois, flew communication and reconnaissance missions in support of the expedition. Communication missions included flying dispatches and mail to forward units, as well as finding forward units out of radio contact. It also, unexpectedly, provided ground logistical support because the squadron possessed organic transport to provide for its own needs. On multiple occasions a squadron aircraft flying dispatches to a forward unit was informed of severe supply shortages, which the squadron remedied using its own ground vehicles to transport the supplies. This first operational employment of the squadron was an exceptionally demanding test. Rather than operating from a fixed airfield, the squadron had to follow Pershing's columns as they advanced deep into Mexico.[14]

---

[10] Maurer Maurer, *The U.S. Air Service in World War I:* Vol. II, *Early Concepts of Military Aviation*, Washington, D.C.: Office of Air Force History, 1978b, pp. 1, 3, 19.

[11] Miller, 2003, p. 5.

[12] U.S. Air Force, *A Guide to United States Air Force Lineage and Honors*, Maxwell Air Force Base, Ala.: U.S. Air Force Historical Research Agency, undated.

[13] U.S. Air Force, *Air Force Instruction 38-101, Air Force Organization*, Washington, D.C.: Department of the Air Force, March 16, 2011a, pp. 13–14.

[14] Miller, 2003, pp. 12, 18–19.

Foulois, his officers, and his enlisted men had no logistics infrastructure, policies, or procedures to follow in this power projection mission. Having already constructed their own facilities at multiple locations and done a test road deployment from Fort Sill, Oklahoma, to Fort Sam Houston, Texas, squadron members continued to invent new support concepts and capabilities, such as a mobile machine shop. Despite the energy and innovative spirit of squadron members, the Curtiss JN-2 aircraft they flew were inadequate for the demands of the Mexican expedition. The JN-2s were grossly underpowered, limiting payload and, most important, the ability to fly at high altitudes and in high winds, both of which were necessary in Mexico. Also, despite the best efforts of the squadron maintenance personnel, the aircraft of that time were not reliable. Instead of eight aircraft, the squadron needed several dozen to sustain operations. Indeed, European air forces were finding that they needed 36 aircraft in a squadron (12 operating, 12 replacement, and 12 reserve). After only a month of operational service, the squadron has lost six of its eight aircraft and was forced to return to Columbus, New Mexico, to refit and test alternative aircraft designs.[15]

The poor performance of the aircraft (including shoddy quality control at Curtiss and other manufacturers) became public knowledge and became a bit of a scandal for the Army and the War Department. The negative press coverage and public outcry ultimately had a positive effect, however, triggering a massive infusion of funding from Congress. The almost $14 million that Congress authorized for Army aviation following the expedition was nine times the funding allocated in prior years for American military aviation.[16]

Why were the first American flying units organized into squadrons as opposed to battalions or some other formation?

The term *squadron* was first coined by the Italians in the mid-sixteenth century to describe a square tactical formation of soldiers, the *squadrone* (from *squadra*, the Italian word for "square").[17] *Squadron* was commonly used to describe a tactical organization from that point on. As John Wilson observes, "At the beginning of the seventeenth century armies had no permanent tactical subdivisions. Administrative organizations called "regiments" were primarily designed to bring armed men to the battlefield. Upon arriving at the battle site, the men were usually grouped into battalions or squadrons, tactical organizations."[18] *Squadron* was also used to designate naval units as early as 1588. In the United States its use had become common in the Navy by 1801 when the Mediterranean Squadron was deployed for

---

[15] Miller, 2003, pp. 6, 14, 41.

[16] Alfred F. Hurley, *Billy Mitchell: Crusader for Air Power*, Bloomington: Indiana University Press, 1975, p. 20.

[17] Oxford University Press, "Squadron," in *Oxford English Dictionary Online*.

[18] John B. Wilson, *Maneuver and Firepower: The Evolution of Divisions and Separate Brigades*, Washington, D.C.: U.S. Army Center of Military History, 1998, p. vii.

the First Barbary War.[19] Finally, by the mid-nineteenth century, U.S. Army cavalry units had settled on the squadron as the battalion-level unit of organization. Battalions were composed of two or more elements called troops.[20]

By 1913, when the 1st Aero Squadron was created, there was a long history of using this terminology for a variety of military formations. The official USAF explanation for the choice of the term *squadron* suggests the Army was influenced by doctrine, as well as the precedent set by older aviation organizations. First, both American and French military doctrine saw the airplane as "an extension of the cavalry—in effect a sky cavalry." U.S. cavalry units were organized into squadrons; it was reasonable that aviation would be similarly organized. Second, and likely the real driving force behind the decision, the Royal Flying Corps had created its No. 1 Squadron in May 1912; after that, "other nations quickly followed the British example."[21]

The squadron quickly took hold as the basic formation for Army aviation. For example, a 1915 Army War College report recommended an Army organization of five tactical divisions and eight aero squadrons: one for each division, plus a squadron for each overseas garrison in Hawaii, Panama, and the Philippines.[22] In 1917 the War Department published new tables of organization for infantry and cavalry divisions. This new structure included an aero squadron for battlefield "reconnaissance and observation" in both the infantry and cavalry divisions. The squadron was equipped with 12 aircraft.[23]

World War I

The start of World War I in August 1914 had remarkably little immediate effect on U.S. Army aviation force structure. Even as late as March 1917 (a month before the United States declared war on the German Empire) the Army possessed only five aero squadrons. By the end of May, this force had doubled to ten, but it remained a tiny and poorly equipped air force compared to those of the other combatants.[24]

---

[19] Ian Toll, *Six Frigates: The Epic History of the Founding of the U.S. Navy*, New York: W. W. Norton and Company, 2008, p. 169. See also Oxford University Press, "Squadron," for a 1588 reference to a "squadron of Galeons of Portugall."

[20] U.S. Air Force, *A Guide*, undated; Wilson, 1998, p. 21, footnote 35.

[21] U.S. Air Force, *A Guide*, undated; Royal Air Force, *1 (F) Squadron*, undated.

[22] David E. Johnson, *Fast Tanks and Heavy Bombers: Innovation in the U.S. Army, 1917–1945*, Ithaca, N.Y.: Cornell University Press, 1998, p. 41.

[23] Wilson, 1998, p. 38.

[24] Daniel Haulman, *Table of Ten Oldest USAF Squadrons*, Maxwell Air Force Base, Ala.: Air Force Historical Research Agency, February 3, 2016. Regarding the poor state of military aviation in the United States in 1917, see Herbert A. Johnson, *Wingless Eagle: U.S. Army Aviation Through World War I*, Chapel Hill: University of North Carolina Press, 2001.

U.S. Air Service units (under the American Expeditionary Force) began deploying to France as squadrons in 1918. The first unit assigned to the front was the 1st Corps Observation Squadron, arriving on April 8, 1918.

The soon to be famous 94th Aero Squadron arrived at the front a day later.[25] Following the British example, U.S. squadrons were organized into pursuit, bombardment, and surveillance groups.[26] Of these, only bombardment groups envisioned flying all their aircraft together on raids.[27] The group was, however, an important command element for pursuit squadrons. Group officers were responsible for adjutant, supply, transportation, operations, intelligence, photographic, and communications functions.[28] With regard to operations, the 1st Pursuit Group operations order for October 22, 1918, makes clear that it was the group, not the squadron, that determined which missions would be flown, and by whom, on a given day. Operations Order Number 41 detailed how many aircraft would be on alert, at what times, and from which subordinate squadron. It also dictated patrol schedules, routes, and altitudes.[29]

According to the chief of Air Service, 1st Army, bombardment groups were composed of four squadrons with a total of 100 aircraft. Missions could be conducted at the squadron or group level. Pursuit groups also contained four squadrons, but the flight was the tactical unit of employment for fighter aircraft as opposed to the squadron and group for bombardment units. Each of the three flights in a pursuit squadron typically flew independent missions.[30] A World War I airfield usually hosted one group of four squadrons with 80 to 90 aircraft and roughly 1,000 personnel.[31]

The aviation wing was also created during this war. The 1st Army Air Service under COL William "Billy" Mitchell included the 1st Pursuit Wing, the Night Bombardment Wing (French and Italian units), and the Corps Observation Wing.[32] General Orders Number 8 specified that

---

[25] Maurer Maurer, ed., *The U.S. Air Service in World War I:* Vol. I, *The Final Report and A Tactical History*, Washington, D.C.: Office of Air Force History, 1978a, p. 18.

[26] James J. Hudson, *Hostile Skies: A Combat History of the American Air Service in World War I*, Syracuse, N.Y.: Syracuse University Press, 1968, pp. 63–64.

[27] U.S. Air Force, *A Guide*, undated.

[28] Maurer, 1978a, pp. 173–174.

[29] The full order is reproduced in the USAF official history of the Air Service in World War I. See Maurer, 1978a, Appendix C, pp. 348–349.

[30] William Mitchell, *Our Air Force: The Keystone of National Defense*, New York: E. P. Dutton and Company, 1921, pp. 52, 57. See also William Mitchell, *Memoirs of World War I: From Start to Finish of Our Greatest War*, New York: Random House, 1960.

[31] Eddie Rickenbacker, *Fighting the Flying Circus*, New York: Doubleday, 1965, pp. 9–10.

[32] Maurer, 1978b, p. 237.

a wing in the Air Service "consists of two or more groups; a group consists of two or more squadrons." It also described the duties of wing commanders:

> The Commander, Pursuit Wing under the C.A.S. [Chief, Air Service], 1st Army, is in direct command of all pursuit groups composing the wing and is responsible for the proper execution of the duties of each group. He will assure proper telephonic, telegraphic and radio communication and liaison between his command and other arms or services of the Army. He will exercise technical supervision over all material and personnel in the wing and will see that all missions assigned are promptly and efficiently carried out. He is responsible that all records, reports, messages and orders are properly made out, filed and forwarded and that the methods of handling and transmitting information work efficiently.[33]

General Orders Number 8 went on to observe that the duties of Army observation and bombardment wing commanders were similar. It also specified group commander duties:

> The Group Commander, Corps Observation, Pursuit or Bombardment commands his group, and is responsible that all orders for missions from higher authority are promptly executed. He assigns specific duties to each squadron, is responsible that Squadron Commanders understand their missions, and that all group specialties such as photographic sections, operations offices and working with Branch Intelligence Officers are properly kept up and executed. He is responsible for the proper preparation, transmission, or filing of all reports, messages and orders to their destination, and that all personnel and material is in condition for effective work.[34]

Operations orders originated in the G-3 of 1st Army Headquarters, American Expeditionary Force. The orders were sent to Colonel Mitchell as chief of Air Service, First Army. Mitchell would then develop and send a plan of employment to the wings (or independent groups, as appropriate). The wings, in turn, would then assign responsibilities to their groups.[35] As noted above, the groups would further specify squadron responsibilities.

Before the war ended, the Air Service had grown in competence to the point that it could conduct multigroup operations. For example, during the Saint-Mihiel Offensive of September–October 1918, Mitchell employed two pursuit groups and one group of day bombers. As Wesley Frank Craven and James Lea Cate note, "These struck at troop concentrations and communications and attacked airdromes behind enemy lines with the purpose of destroying

---

[33] Reproduction of General Orders Number 8, Headquarters 1st Army, American Expeditionary Forces, France, August 10, 1918 in Maurer Maurer, ed., *The U.S. Air Service in World War I:* Vol. III, *The Battle of St. Mihiel*, Washington, D.C.: Office of Air Force History, 1978c, pp. 9–10.

[34] Maurer, 1978c, pp. 9–10.

[35] Maurer, 1978b, pp. 231–249.

the enemy's installations and planes on the ground or forcing him to come up to fight at a disadvantage."[36]

Although organized at the front into groups and wings, the squadron remained the force-sizing metric for Army aviation throughout the war. This is illustrated by the seven plans to grow Army aviation that were developed, debated, and discarded between 1917 and 1918. The first envisioned 59 squadrons; other plans called for between 120 and 358 squadrons. The final "202 program" plan of August 1918 called for 202 squadrons to be deployed at the front by July 1919.[37] The war ended long before this ambitious plan could be realized.

Figure 2.1 illustrates the actual growth of the Army Air Service (AAS) measured in frontline squadrons from April 1918 to Armistice Day on November 11, 1918.[38]

Figure 2.1. Cumulative U.S. Air Service Squadrons Deployed to the Front, April–November 1918

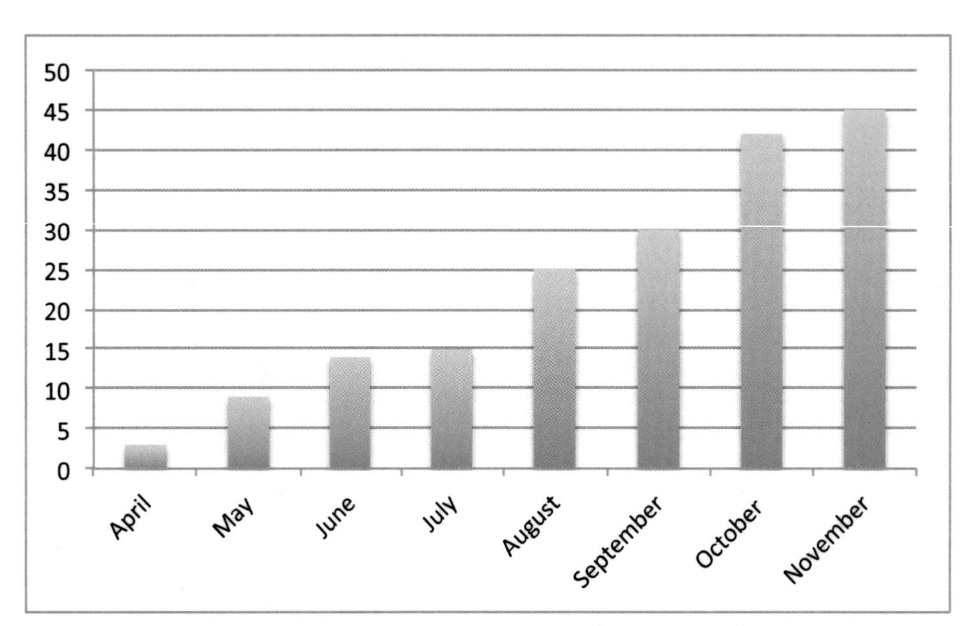

SOURCE: RAND analysis based on data from Maurer Maurer, ed., *The U.S. Air Service in World War I:* Vol. I, *The Final Report and a Tactical History*, Washington, D.C.: Office of Air Force History, 1978a, p. 18.

---

[36] Wesley Frank Craven and James Lea Cate, *The Army Air Forces in World War II:* Vol. 1, *Plans and Early Operations, January 1939 to August 1942*, Washington, D.C.: Office of Air Force History, 1983a, pp. 14–15.

[37] Maurer, 1978a, pp. 127–129, 135–137, 167–171, 195–196, 219–220, and 227–228; Maurer's Appendix B, p. 412, summarizes the key features of the seven plans. See also Mason M. Patrick, *The United States in the Air*, Garden City, N.Y.: Doubleday, Doran and Company, 1928, p. 17.

[38] These 45 squadrons included 18 observation, 20 pursuit, and seven bombardment. See Johnson, 1998, p. 46; and Hudson, 1968, p. 7. Note that in August 1918, General Pershing and the War Department had agreed on an ambitious plan to greatly expand the AAS to 202 aero squadrons by July 1919. Of these planned squadrons, 60 would be pursuit, 101 observation, and 41 bombardment. See Hudson, 1968, p. 7.

What should we conclude from this brief survey about force presentation in World War I? Did the AAS present forces at the squadron or group level? The case for the squadron is twofold. First, as noted above, forces were deployed to the front as squadrons. Second, the Army and the War Department planned the force in terms of squadrons, not groups or wings. The argument for the group as the force presentation construct is that it commanded the squadrons and provided vital enabling functions. The group ran the airfield, directed squadron operations, and provided critical support without which the squadrons would have limited combat power. Had the war continued into 1919, the group (and wing) would have become even more important. Mitchell and other air planners envisioned massive, multigroup bombing raids. To control such large formations, a higher level tactical organization (e.g., a wing or air division) would be needed as would later be done during the largest bombing raids in World War II. In this sense, the AAS presented forces in a manner remarkably similar to the USAF today. Based on this review of the World War I record, we have to conclude that the AAS presented forces at the squadron level for force sizing and deployment, it employed forces at the squadron and group level, and it sustained operations at the group level.

The Interwar Period

As of December 1918, the American Air Service included 15 organizations above the squadron level: one pursuit wing and 14 groups.[39] The groups and wing disbanded overseas while squadrons and balloon companies returned to the United States. Although most squadrons were disbanded, twelve survived. In 1919 these squadrons were organized into pursuit, bombardment, and surveillance groups (four squadrons per group) under a single wing structure. This composite wing was based at Kelly Field, Texas, along the still restive Mexican border. By 1923, Army aviation consisted of one wing, three combat groups (one pursuit, one bombardment, one attack), one group headquarters, and 11 observation squadrons in the United States. Each combat group contained four squadrons. Additionally, composite groups were based overseas in Hawaii, Panama, and the Philippines. Each was assigned one group with three squadrons (one observation, one pursuit, one bombardment). Hawaii was assigned an extra group headquarters and two additional squadrons.[40]

Although not directly salient to our exploration of force presentation, it is hard to avoid entirely the long and bitter struggle over air force independence. Beginning with the return of Billy Mitchell from Europe, prominent airmen fought for an independent service. For some, like Mitchell, airpower's full potential as a strategic instrument could only be realized through a separate branch dedicated to advancing the air instrument. For others, like GEN Mason

---

[39] See U.S. Air Force, *A Guide*, undated.

[40] Maurer Maurer, *Aviation in the U.S. Army, 1919–1939*, Washington, D.C.: Office of Air Force History, 1987, pp. 4, 70–71.

Patrick, some level of separate organization (e.g., along the lines of the Marine Corps within the Department of the Navy) was necessary to ensure that funding, equipment, and training were not neglected by big Army needs.[41] The back and forth over airpower led to a series of incremental steps forward. First came the creation of the Army Air Service in 1918, then the Air Corps in 1926.

The next step in the gradual evolution toward a fully independent air force came on March 1, 1935, when the General Headquarters Air Force (GHQ Air Force) was created, placing all tactical aviation units under one headquarters (except for observation squadrons assigned to ground forces).[42] Air Force historian Maurer Maurer describes the complex relationship between the GHQ and the Air Corps:

> As Commanding General, GHQ Air Force, [BG Frank] Andrews carried responsibility for the instruction, training, maneuvers, and tactical employment of all elements of his command. Since the Air Corps still handled individual training and materiel, it supplied officers, enlisted men, and equipment to the Air Force.[43]

There were three wings and one airship (lighter than air) group in the GHQ. The wings commanded nine groups with a total of 30 squadrons. All were composite organizations, mixing bombardment, attack, and pursuit groups. Additionally, observation squadrons were attached to the 1st and 2nd Wings.[44] Writing shortly before the start of World War II, two prominent airmen described Air Corps organization as follows:

- The *wing* "is a tactical air unit composed of two or three groups . . . . It is envisaged as the largest air fighting unit which one commander can efficiently control and directly supervise. It is a tactical command as differentiated from an administrative command."
- The *group* "is normally composed of three squadrons. It was conceived as the largest air unit which one leader can efficiently control in the air."
- The *squadron* "is deemed essential, as is the battalion in the Infantry, in order to have a unit sufficiently small in size to receive personal supervision, direction and control of one experienced officer, and in order to provide for detailed training, and firsthand direction of supply, discipline and combat methods."[45]

The group continued to be the force-sizing metric for the Air Corps as World War II drew near. By the spring of 1939, the Air Corps developed a plan that would field 24 combat-ready

---

[41] Robert P. White, *Mason Patrick and the Fight for Air Service Independence*, Washington, D.C.: Smithsonian Institution Press, 2001, p. 6.

[42] Maurer, 1987, p. 328.

[43] Maurer, 1987, p. 328.

[44] Maurer, 1987, pp. 327–328.

[45] H. H. Arnold and Ira C. Eaker, *Winged Warfare*, New York: Harper and Brothers, 1941, pp. 92–94.

groups by June 1941. By 1940 international events had overtaken the modest 1939 plan, replacing it with the 41-group plan. A short time later this objective was expanded to 54 groups and 4,000 combat aircraft. By 1941 this plan also proved inadequate and was replaced with a plan to deploy 84 groups by 1942. None of these plans was fully executed prior to the Japanese attack on Pearl Harbor. On December 7, 1941, 70 groups had been activated but "many of the units were at cadre strength only, and few had been equipped with suitable aircraft."[46] In fact, none of these plans was remotely close to what the United States would need or field during the War. This is in contrast to the exceptionally accurate estimates made by LTC Harold George and his Air War Plans Division staff. George, who was working with LTC Kenneth Walker, Maj Laurence S. Kuter, and Maj Haywood S. Hansell, Jr. (all soon to be famous airmen), "declared that the Air Corps would need 201 combat groups without the proposed B-36 type bomber and 251 with the B-36."[47]

World War II

The scale and diversity of American air operations during World War II is so great that even a focus on force presentation is too large a canvas. Additional bounding is required; this section limits the analysis to four aspects of force presentation: (1) its use for force sizing, (2) force presentation innovations during the height of the U.S. Army Air Forces (USAAF) strategic bombing of Germany (1943–1945), (3) tactical air operations in the South and Southwest Pacific (1942–1945), and (4) B-29 operations against the Japanese homeland (1944–1945). Space and time constraints prevent the inclusion of operations in and from North Africa or tactical air operations in Europe between 1944 and 1945.

As noted in the previous section, the Air War Plans Division had estimated that the war would ultimately require 201 combat groups, almost three times the prewar force of 70 understrength groups. Once the war began, the new Army Air Forces developed plans for a dramatic expansion in air combat capabilities. In January 1942 the 115-group plan was approved. By July 1942 the goal had expanded to 224 groups, and to 273 groups by September of that year. The actual wartime peak was 243 groups in April 1945, with 224 of those groups deployed abroad.[48] Table 2.1 gives the breakdown of combat groups by aircraft category.

---

[46] Craven and Cate, 1983a, p. 105.

[47] James P. Tate, *The Army and Its Air Corps: Army Policy Toward Aviation, 1919–1941*, Maxwell Air Force Base, Ala.: Air University Press, 1998, p. 174.

[48] Wesley Frank Craven and James Lea Cate, *The Army Air Forces in World War II:* Vol. VI, *Men and Planes*, Washington, D.C.: Office of Air Force History, 1983d, p. 32.

Table 2.1. World War II Peak Number and Types of Combat Groups

Type	Number of Groups
Bombardment	127
Fighter	71
Troop Carrier	32
Reconnaissance	13
Total	243

SOURCE: RAND analysis based on data from Wesley Frank Craven and James Lea Cate, *The Army Air Forces in World War II:* Vol. VI, *Men and Planes*, Washington, D.C.: Office of Air Force History, 1983d, p. 32.

In their history of the AAF in World War II Craven and Cate describe its organization in action. They note that the aircrew was the basic level of organization, whether a single pilot in a pursuit aircraft or "eleven men working together in a B-29 bomber." Aircraft were typically organized into flights of three or more aircraft, under a flight lead. Combat squadrons typically had 200 to 500 men and seven to 25 aircraft (plus reserves) for bomber and fighter units, respectively. Squadrons were typically assigned three or four to a group but often operated independently. "The group, which may be considered as roughly parallel to an Army regiment, was the key unit for administrative and operational purposes . . . . By 1945 a group normally had a personnel strength of about 990 officers and men for a single-engine fighter group, or 2,260 for a heavy bomber group."[49]

The wing played an ambiguous role in World War II, at times vital but at other times secondary to air divisions and functional commands. For example, the XIX Tactical Air Command—the air component teamed with GEN George Patton in the 1944 sprint for the Rhine—concluded during the campaign in France that wings were "an unnecessary administrative echelon and recommended their elimination in future operations." In at least one case, a fighter group conducted combat operations for four months directly under a tactical air command while its wing headquarters was still in the United States.[50] Craven and Cate

---

[49] Craven and Cate, 1983d, pp. 58–59.

[50] David N. Spires, *Patton's Air Force: Forging a Legendary Air-Ground Team*, Washington, D.C.: Smithsonian Institution Press, 2002, pp. 37–38.

describe the wing's transition from the key tactical and administrative command during the interwar years to a bit of a fifth wheel during World War II:

> Before the war the wing had served as the key tactical and administrative organization through which the GHQ Air Force directed its combat forces. The wing continued to have some utility during the war, primarily for purposes of tactical control, but the functionally conceived command, whose development seems to reflect another influence of RAF patterns on AAF organization, came to be the chief agency for coordination of effort between a top air commander and the groups making up his force.[51]

Yet in another volume of their history, Craven and Cate make an argument for the unique value of the wing, describing the contributions of the 1st Bombardment Wing in identifying and disseminating best practices for strategic bombing of heavily defended targets in Germany.

> When General [Laurence S.] Kuter took over the wing on 6 December 1942, he found the four groups each operating according to its own tactical doctrine. No wing organization existed for tactical purposes, and consequently the groups collaborated only in the sense that they all attacked the same target roughly at the same time. No effort was made to secure additional fire support by coordinating group tactics. Squadrons and groups had developed into cohesive teams, but the wing as a whole had not become a combat unit.[52]

General Kuter believed that the larger the formation, the more mutual defensive support could be provided. He thus sought to bring squadrons and groups into the largest formations practical. He invented a tactic using 18 to 21 bombers that was called the combat box. This became the standard combat formation; it was the smallest number of aircraft that could provide mutual support, while at the same time it was the largest formation that could be "handled readily on the bombing run." During the flight to and from the target, the 1st Bombardment Wing found that two to three combat boxes were the largest practical formation. They also found that two groups were the largest that could be briefed and controlled by a single commander. As a result, the 1st Bombardment Wing created two combat wings of two groups each under the command of the senior group commander, who "was given full authority in planning and executing the mission. This organization existed for tactical purposes only and in no way affected the administrative organization of the bombardment wing."[53]

The use of combat wings as temporary tactical organizations for specific missions became the standard in the VIII Bomber Command (later the 8th Air Force), used in some of the most

---

[51] Craven and Cate, 1983d, p. 59.

[52] Wesley Frank Craven and James Lea Cate, *The Army Air Forces in World War II:* Vol. II, *Europe: Torch to Pointblank*, Washington, D.C.: Office of Air Force History, 1983b, p. 264.

[53] Craven and Cate, 1983b, p. 267.

famous missions, such as the Schweinfurt-Regensburg raid of August 17, 1943.[54] Although the group, air division, and functional command (e.g., the XIII Fighter Command) may have supplanted some of the duties of the wing, the wing echelon of command remained in use throughout the war in all theaters.

Since the bombers had to fly in combat boxes for self-protection and in large numbers to mass effects on targets, force employment against targets in Germany was typically at the wing/air division level rather than the squadron/group level. For example, Operation Argument began on February 19, 1944, with 16 combat wings of bombers (over 1,000 bombers) supported by 17 groups of fighters.[55]

With such large formations, the orchestration of the rendezvous became high art. The success of bombing missions was heavily dependent on the skill of squadron, group, and wing navigators who managed the complex calculations to rendezvous on time and in a safe manner. With hundreds of aircraft launching from dozens of bases—often in the dark and/or in poor weather—this called for exceptional skill. Squadrons had to rendezvous to form groups, in groups to form combat wings, and in wings to form air divisions.[56] Navigator Harry Crosby describes multiple missions where the entire 3rd Air Division flew, including an April 11, 1945, raid on Regensburg that involved 1,300 aircraft.[57]

In the Pacific Theater, two campaigns—tactical air operations in the South/Southwest Pacific and strategic bombing of the Japanese homeland—illustrate how force presentation constructs varied to meet diverse operational demands.

The first campaign, in the South and Southwest Pacific, lasted from 1942 to 1945. It was characterized by (1) wide dispersion of units, (2) squadrons and groups operating far forward of their higher commands, (3) primitive and often temporary forward fields, (4) limited navigation aids and weather information, (5) long overwater missions, (6) severe tropical weather, and (7) a wide range of missions (including maritime interdiction). The 13th Air Force, which conducted many of these operations, was rightly known as "the Jungle Air Force." Unlike Europe, however, there were no massive bombing raids against industrial or

---

[54] For a detailed treatment of the twin raids, see Thomas M. Coffey, *Decision over Schweinfurt: The U.S. 8th Air Force Battle for Daylight Bombing*, New York: David McKay Company, 1977.

[55] Mark Clodfelter, *Beneficial Bombing: The Progressive Foundations of American Air Power, 1917–1945*, Lincoln: University of Nebraska Press, 2010, p. 157.

[56] Navigation to the target was the responsibility of the mission lead navigator. For a detailed discussion of the role of navigators in strategic bombing, see Harry H. Crosby, *A Wing and a Prayer: The "Bloody 100th" Bomb Group of the U.S. Eighth Air Force in Action over Europe in World War II*, New York: iUniverse.com, Inc., 2000.

[57] Crosby, 2000, p. 369.

urban targets. Nor were there the sophisticated and dense air defenses that the 8th Air Force faced over Germany. As Craven and Cate observe,

> Bombardment by the several Army air forces in the Pacific . . . had been almost exclusively tactical, directed against the enemy's air strips, at the shipping whereby he nourishes his advanced forces, at his supply dumps and island defenses, against his troops in the field.[58]

Weather and long flights over open water were often as much of a threat as Japanese air defenses. USAAF tactics and organizations reflected these unique operational and tactical demands. With smaller units operating in smaller packages against less dense air defenses than was typical in Europe, tactics such as the combat box and mission organizations like the combat wing were not necessary. The standard practice in the South Pacific was for squadrons and groups to operate from forward bases that were often primitive and temporary. The numbered air forces (the 5th and 13th) were based at relatively more advanced facilities (initially at Noumea, New Caledonia, and Brisbane, Australia). Each numbered air force had a subordinate command responsible for bomber or fighter operations. These were the XIII Bomber and Fighter Commands and V Bomber and Fighter Commands, respectively. Like the numbered air forces, their subordinate commands were typically in rear areas. Consequently, the group was the largest tactical organization for pursuit and bomber aircraft operating in the South Pacific.[59]

The second campaign, from June 1944 to August 1945, was conducted by B-29 bombers against urban and industrial targets in the Japanese homeland. The B-29 was a new aircraft with significantly greater capabilities than other U.S. bombers; thus, wing-level organizations were created in the United States to train crews on its use. The first of these, the 58th Bomb Wing was established in Georgia in June 1943; it arrived in India in the spring of 1944.[60] Two bomber commands (the XX and XXI) were created in 1944 to oversee the air campaign with six subordinate wings and a total of 24 bomber groups.

The first mission by elements of the 58th Bomb Wing (under the XX Bomber Command) struck railroad yards in Bangkok, Thailand, but the main targets were in the Japanese homeland. Even with forward bases in China, the B-29s could only reach the southernmost of the Japanese islands, but they were able to strike steel plants in Kyusha, marking the beginning of a more classic strategic air campaign. The India-based campaign against Japan was extremely

---

[58] Wesley Frank Craven and James Lea Cate, *The Army Air Forces in World War II*: Volume V, *The Pacific: Matterhorn to Nagasaki*, Washington, D.C.: Office of Air Force History, 1983c, p. 3.

[59] The most detailed treatment of air warfare in the South Pacific is Eric M. Bergerud, *Fire in the Sky: The Air War in the South Pacific*, Boulder, Colo.: Westview Press, 2001.

[60] Kenneth P. Werrell, *Blankets of Fire: U.S. Bombers over Japan During World War II*, Washington, D.C.: Smithsonian Institution Press, 1996, p. 91; Daniel Haulman, *Hitting Home: The Air Offensive Against Japan*, Washington, D.C.: Air Force History and Museums Program, 1999, p. 9. See also Craven and Cate, 1983c, Chapters 1 and 4.

challenging, requiring all fuel and supplies for forward bases in China to be airlifted over the Himalayas, at great cost in men and aircraft.[61] Operations against Japan from India were discontinued as soon as alternative bases in the Pacific became available. As Daniel Haulman notes, "By late 1944, American bombers were raiding Japan from the recently captured Marianas, making operations from the vulnerable and logistically impractical China bases unnecessary. In January 1945, the XX Bomber Command abandoned its bases in China and concentrated 58th Bomb Wing resources in India."[62] The 58th continued to operate from India until March 1945, when it moved to the Marianas. Between January and March it supported British operations in Burma by attacking airfields, ports, railroads, and oil refineries in Burma, the East Indies, Indochina, Malaya, Singapore, and Thailand.[63]

In November 1944 the 73rd Bomb Wing (XXI Bomber Command) began striking Japan with B-29s launching from airfields on Saipan, Tinian, and Guam in the Marianas. The shorter range made more frequent and larger attacks possible, but initial results were disappointing. High winds and a more distributed target set made high-altitude precision bombing less effective than it had been against Germany. The scale and intensity of attacks increased greatly, however, after Maj Gen Curtis LeMay assumed command of the XXI Bomber Command in January 1945. LeMay, at the urging of Washington, began experimenting with incendiaries, discovering them to be effective against densely packed wood structures typical of Japanese urban centers. This firebombing campaign is most famous for the March 9, 1945, attack on Tokyo. The 344 bombers from the XXI Bomber Command attacked the core of Tokyo, causing a firestorm that destroyed 270,000 buildings and killed 84,000 people.[64] The campaign was then expanded to include other urban and industrial centers.

Between November 1994 and July 1945, the XXI Bomber Command would grow to a total of five B-29 wings operating from their own airfields on Guam (two), Saipan (one), and Tinian (two).[65] This suggests that these wings had responsibility for overseeing the operation of their respective airfields, foreshadowing the role that the wing plays today as air base operator. Additionally, in April 1945, fighter groups began flying escort missions from Iwo Jima, and by summer 1945 the 301st Fighter Wing was operating out of Ie Shima, Okinawa.[66]

---

[61] See William H. Tunner, *Over the Hump*, Washington, D.C.: Air Force History and Museum Program, 1998, pp. 43–135; Otha C. Spencer, *Flying the Hump: Memories of an Air War*, College Station: Texas A&M University Press, 1992.

[62] Haulman, 1999, p. 12.

[63] Haulman, 1999, p. 12.

[64] Craven and Cate, 1983c, pp. 569–570, 573. For more details of the firebombing campaign see Werrell, 1996.

[65] Haulman, 1999, p. 19; Werrell, 1996, pp. 101–103, 254–255.

[66] Ie Shima (now called Ie Jima) is an island just northwest of Okinawa; it is part of Okinawa Prefecture. See Craven and Cate, 1983c, p. 655.

What are we to make of force presentation in World War II? It should be no surprise that in a conflict of such scale and intensity there was no single, uniform approach for the organization, training, deployment, employment, and command of air forces. The squadron, group, wing, air division, numbered command, and numbered air force all played important roles, depending on the time and place. As in World War I, the squadron remained the one constant, although typically operating as part of a combat group. The group was arguably the single most important echelon of command across theaters and over the course of the conflict. If one had to select a single force presentation construct for the employment of airpower in World War II, the group would have to be it. It also remained the force-sizing metric for the USAAF through the end of World War II and for several years beyond. That said, as the conflict reached its climax in 1944–1945, the sheer number of aircraft involved necessitated a larger role for higher echelons of command, including the wing, air division, functional command, and numbered air force.

Chapter Three begins with a discussion of the organizational challenges facing the USAF as the newly independent service sought to integrate jet aircraft and nuclear weapons into its concepts and organizations.

3. USAAF and USAF Force Presentation Constructs from 1946 to 2016

This chapter considers the evolution of USAF force presentation constructs from the end of World War II to 2016. It concludes with an analysis of historical trends in force presentation.

The Transition to Peace

Formal planning for the postwar Air Force began several years before the end of World War II. In a July 29, 1943, memo to Army Chief of Staff George C. Marshall, Assistant Chief of Staff MG Thomas Handy laid out a postwar plan for a U.S. Army (including the Army Air Forces) of roughly 1.5 million men, 28 divisions, and 105 air groups.[1] General Marshall, however, did not believe that Congress would appropriate sufficient funds to support such a large standing army. These concerns led to dramatic reductions in the planned force. The February 1945 plan approved by Marshall envisioned an Army Air Force of only 16 air groups. Senior airmen considered this force the equivalent of unilateral disarmament and according to Sherry, attacked it "for elevating political and budgetary considerations above defense needs."[2] Ultimately, budget and manpower constraints resulted in an August 1945 decision to establish 70 groups as the objective for the permanent force.[3]

Demobilization and budget cuts, however, resulted in a much smaller and less capable force. By December 1946, the USAAF had shrunk to 55 groups, with only two considered combat ready.[4] The size and quality of the force fluctuated over the next five years as the nation debated its postwar defense needs. On the one hand, there was a growing consensus that the nation needed a large and ready air force. Two months after the founding of the U.S. Air Force as a separate branch on September 18, 1947, the report of the President's Air Policy Commission, headed by Thomas Finletter, made a strong case for airpower and recommended a 70-group air force.[5] On the other hand, President Harry S. Truman and his key advisers believed that unchecked defense spending would lead to fiscal ruin. The latter concern

---

[1] Perry McCoy Smith, *The Air Force Plans for Peace: 1943–1945*, Baltimore, Md.: Johns Hopkins University Press, 1970, pp. 54–55.

[2] Michael S. Sherry, *Preparing for the Next War: America Plans for Postwar Defense, 1941–45*, New Haven, Conn.: Yale University Press, 1977, p. 109; see, generally, pp. 101–119.

[3] Smith, 1970, pp. 62–73.

[4] Robert Frank Futrell, *Ideas, Concepts, Doctrine: Basic Thinking in the United States Air Force, 1907–1960*, Maxwell Air Force Base, Ala.: Air University Press, 1989, p. 215.

[5] Thomas K. Finletter, *Survival in the Air Age: A Report*, Washington, D.C.: President's Air Policy Commission, 1947, pp. 24–25.

dominated in 1949 and early 1950, resulting in USAF force cuts that shrunk the force from 55 to 48 groups.

While these force structure and funding controversies raged, the USAF was in the midst of a major reorganization of its tactical units. From 1947 to 1948 the USAF tested a new wing and base organizational structure that for the first time more clearly distinguished between temporary organizations controlled by major commands and permanent organizations controlled by Headquarters Air Force (AFCONs). AFCON units brought together squadrons, groups, and wings into permanent organizations that shared the same numerical designation, lineage, and honors.[6] AFCON combat wings "had a double mission: train for and conduct combat operations (through the combat group and that group's combat squadrons) and operate a permanent installation (through assigned support components)."[7] Each test wing was composed of a combat group, maintenance and supply group, air base group, and station medical group.[8]

Over the next several years, wings slowly replaced groups as the key tactical echelon and force structure began to be measured in wings rather than groups.[9] In 1952 many of the World War II–era combat group headquarters were disbanded. Wing headquarters, particularly those in the Strategic Air Command (SAC) and Air Defense Command (ADC), began to take direct control of subordinate combat and support squadrons. (For more on the SAC, see the next section.) By the mid-1950s, directorates within the wing had replaced all but the air base (combat support) group.[10] Additionally, in 1952 the USAF switched from the group to the wing as the formal force-sizing metric.[11] With the exception of a few independent groups, the combat group disappeared from the USAF for almost forty years.

The debate over USAF force sizing might have gone on for many more years, but the June 1950 invasion of South Korea by forces of the Democratic People's Republic of Korea instantly showcased the serious shortfalls in capability across the entire U.S. military, placing military requirements ahead of fiscal concerns. On September 1, 1950, the Joint Chiefs of Staff approved a USAF buildup to 95 wings by June 1954, a goal that the USAF actually met by June 1952.[12]

---

[6] For example, in the 1st Fighter Wing the subordinate groups would be the 1st Combat Group, 1st Maintenance and Supply Group, etc.

[7] Charles A. Ravenstein, *Air Force Combat Wings: Lineage and Honors Histories, 1947–1977*, Washington, D.C.: Office of Air Force History, 1984, p. xxii.

[8] Craven and Cate, 1983d, p. 59.

[9] The Joint Chiefs of Staff may have preceded the USAF in the use of the wing rather than the group for force sizing. Futrell reports that the Joint Chiefs approved 95 wings in 1950; Ravenstein reports that the USAF switched to the wing for force sizing in 1952. See Futrell, 1989, pp. 317, 327; and Ravenstein, 1984, p. xxii.

[10] U.S. Air Force, *A Guide*, undated.

[11] Ravcnstcin, 1984, p. xxii.

[12] Futrell, 1989, pp. 317, 327.

The Creation and Early Years of the Strategic Air Command

The most consequential organizational change of the early post–World War II years was the creation of the Strategic Air Command. As the only military branch able to deliver nuclear weapons, the Army Air Forces began the postwar era in a unique and institutionally strong position.[13] USAAF leaders recognized the need to create a command dedicated to long-range nuclear and conventional bombing missions.[14] Thus, on March 21, 1946, the Strategic Air Command was created as one of three major combat commands. The other two combat commands were the ADC and the Tactical Air Command (TAC). The Air Material Command, Air Transport Command, Air Training Command, Air University, and USAAF Proving Ground Command were also created as supporting commands. Continental Air Forces was disbanded, with its assets divided among the ADC, SAC, and TAC; its headquarters was redesignated as SAC's.[15]

Carl A. Spaatz, the commanding general of the Army Air Forces, issued SAC's mission statement to Gen George C. Kenney, its first commander:

> The Strategic Air Command will be prepared to conduct long-range offensive operations in any part of the world either independently or in cooperation with land and naval forces; to conduct maximum range reconnaissance over land or sea either independently or in cooperation with land and naval forces; to provide combat units capable of intense and sustained combat operations employing the latest and most advanced weapons; to train units and personnel for the maintenance of the Strategic Forces in all parts of the world; to perform such missions as the Commanding General, Army Air Forces may direct.[16]

This ambitious mission statement embraced the long-held airpower vision of decisive independent bomber operations, but now the force would have global responsibilities and nuclear weapons.[17]

---

[13] This was perhaps best exemplified by President Eisenhower's New Look strategy, which emphasized strategic nuclear capabilities and prioritized airpower investments, leading to the largest USAF budget shares in its history. See Richard M. Leighton, *History of the Office of the Secretary of Defense:* Vol. III, *Strategy, Money and the New Look, 1953–1956*, Washington, D.C.: Historical Office, Office of the Secretary of Defense, 2001.

[14] For more on SAC's role as the first specified command, see Melvin G. Deaile, *The SAC Mentality: The Origins of Organizational Culture in Strategic Air Command, 1946–1962*, dissertation, Chapel Hill: University of North Carolina, 2007, esp. pp. 89–98.

[15] Phillip S. Meilinger, *Bomber: The Formation and Early Years of Strategic Air Command*, Maxwell Air Force Base, Ala.: Air University Press, 2012, p. 77; Futrell, 1989, pp. 206–207.

[16] GEN Carl A. Spaatz, letter, to GEN George C. Kenney, quoted in Meilinger, 2012, p. 77.

[17] For more on the evolution of airpower narratives, see Alan J. Vick, *Proclaiming Airpower: Air Force Narratives and American Public Opinion from 1917 to 2014*, Santa Monica, Calif.: RAND Corporation, RR-1044-AF, 2015, pp. 42–82.

At its creation, SAC inherited a headquarters building at Bolling Field, Washington, D.C., along with 1,300 aircraft, 100,000 personnel, and 52 bases.[18] SAC's tactical order of battle, however, was much smaller, with only 148 B-29s assigned. Figure 3.1 illustrates the growth of SAC's bomber force from 1946 to 1955. In 1946 there were nine very heavy bombardment groups, but only six of these had aircraft assigned. In 1948 SAC transitioned to a mix of heavy bombardment groups (equipped with B-36s) and medium bombardment groups (equipped with B-29s and B-50s).[19] Between 1946 and 1955 SAC grew more than fourfold—from nine understrength bomber groups to 38 full-strength bomber wings.

Figure 3.1. SAC Bomber Groups and Wings, 1946–1955

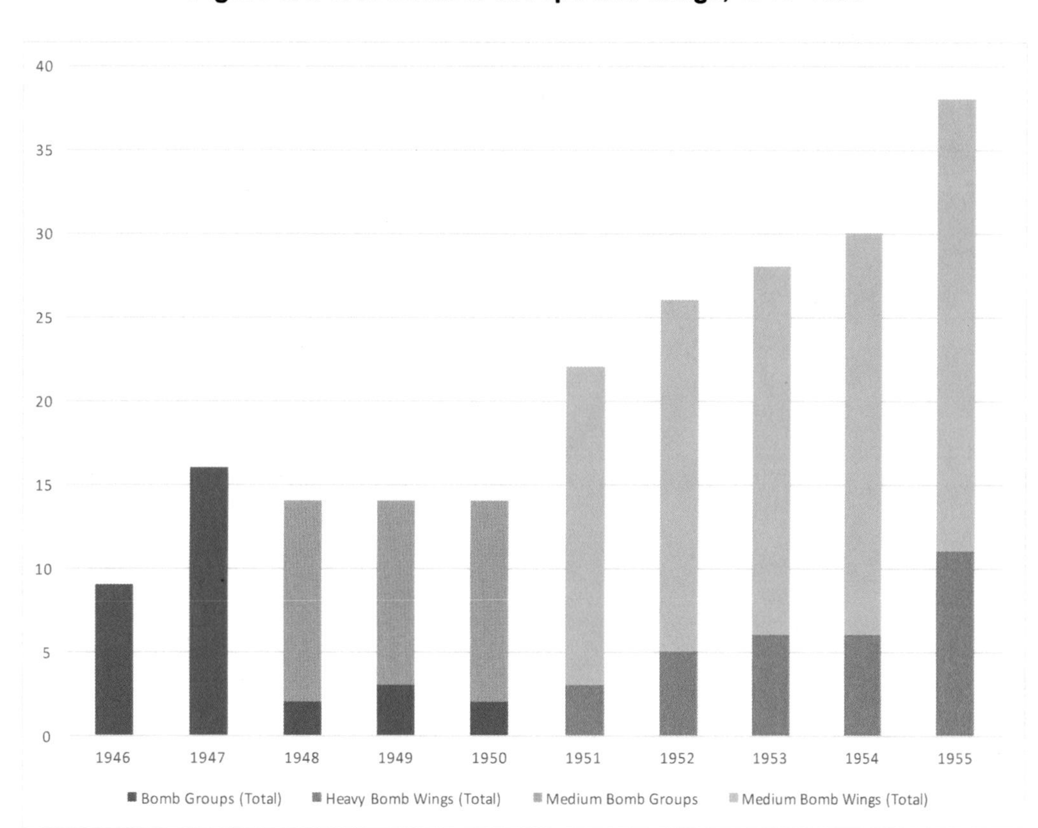

SOURCE: RAND analysis based on data compiled from Norman Polmar, ed., *Strategic Air Command: People, Aircraft, and Missiles*, Annapolis, Md.: Nautical and Aviation Publishing Company of America, 1979, pp. 7–38.

[18] Norman Polmar, ed., *Strategic Air Command: People, Aircraft, and Missiles,* Annapolis, Md.: Nautical and Aviation Publishing Company of America, 1979, p. 7.

[19] Note that B-29s were considered "very heavy bombers" in 1946, but by 1948 they were "medium bombers." See Polmar, 1979, p. 13.

In addition to illustrating the quadrupling in size of SAC between 1946 and 1955, Figure 3.1 also clearly delineates the transition in 1951 from the group to wing structure. Bomber, fighter, and reconnaissance groups were not officially deactivated in SAC until 1952 when all combat squadrons were assigned to the wings but, as Norman Polmar notes, "for all practical purposes, the combat groups had ceased to exist in 1951 when the wings were reorganized and the group headquarters left unmanned."[20] This structure lasted until 1990, when SAC reintroduced groups as part of the USAF-wide objective wing reorganization. As will be discussed later in this chapter, the objective wing created four subordinate groups: operations, logistics, support, and medical. Wing deputies for these functions were replaced by group commanders who oversaw the activities of their subordinate squadrons.

Although the range of SAC's bomber aircraft continued to improve, USAF leaders understood that aerial refueling would be necessary for SAC to become the global force envisioned in its mission statement.[21] Figure 3.2 illustrates the growth of SAC's tanker

Figure 3.2. SAC Tanker Squadrons, 1946–1955

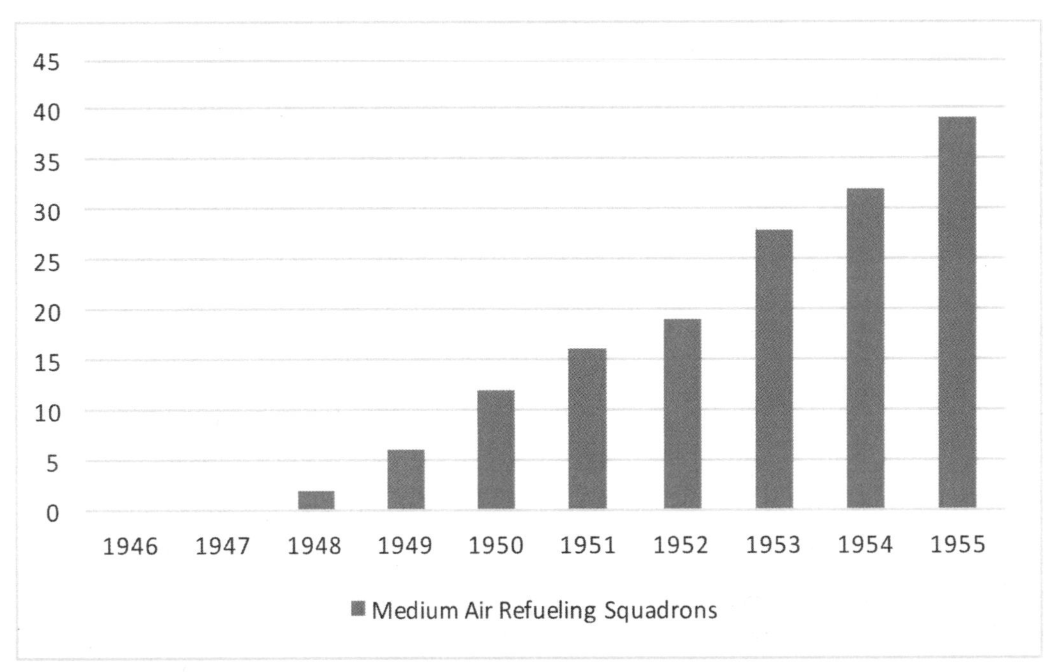

SOURCE: RAND analysis based on data compiled from Norman Polmar, ed., *Strategic Air Command: People, Aircraft, and Missiles*, Annapolis, Md.: Nautical and Aviation Publishing Company of America, 1979, pp. 7–38.

---

[20] Polmar, 1979, p. 28.

[21] In January 1948 Chief of Staff Carl A. Spaatz identified aerial refueling as the USAF's top priority. See Richard K. Smith, *Seventy-Five Years of Inflight Refueling: Highlights, 1923–1998*, Washington, D.C.: Air Force History and Museums Program, 1998, p. 25.

fleet from two medium air refueling squadrons equipped with KB-29 aircraft in 1948 to 39 squadrons equipped with KC-97s in 1955.[22] The first KC-135 tankers became operational two years later in 1957 with the creation of the first heavy air refueling squadrons.[23] Note that, in contrast to bombers, the force-sizing metric for SAC's aerial refueling fleet was the squadron rather than the group or wing.[24]

SAC's rapid growth in force structure demanded a parallel expansion in its basing network, to house its large aircraft inventory not only at home but also at overseas bases to support global operations. As Figure 3.3 shows, SAC's basing network was exclusively in the continental United States (CONUS) for its first four years, varying between 16 and 21 bases. In 1950, one overseas base was added in Puerto Rico. By 1955 there were 14 overseas bases in Guam, North Africa, Puerto Rico, and the United Kingdom.

While SAC was dedicated to preparing for nuclear war, the USAF faced contemporary conventional challenges as well.

Figure 3.3. SAC CONUS and Overseas Bases, 1946–1955

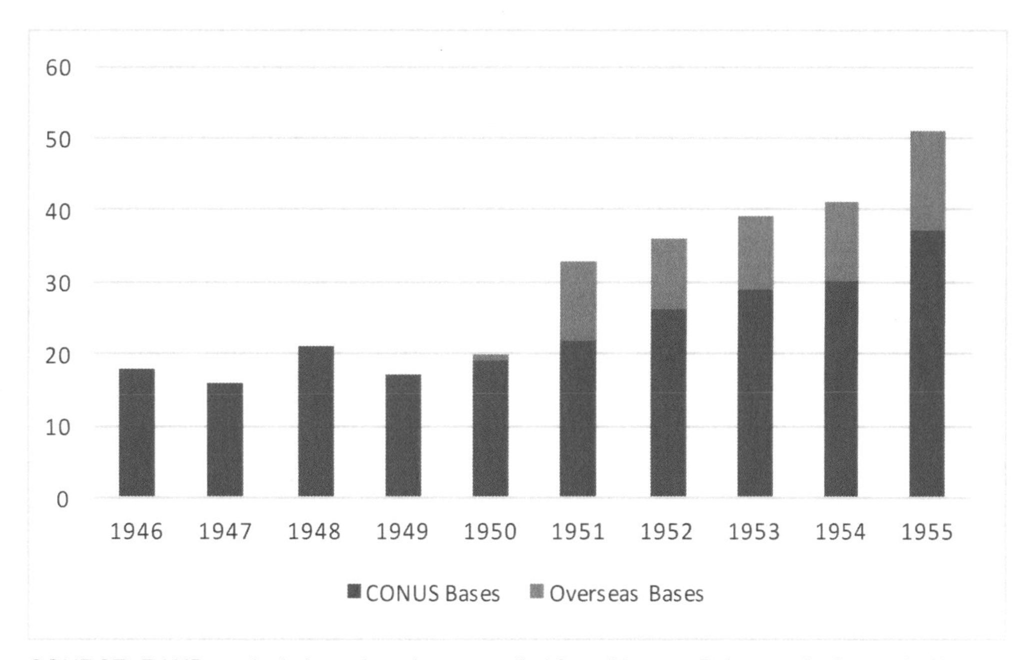

SOURCE: RAND analysis based on data compiled from Norman Polmar, ed., *Strategic Air Command: People, Aircraft, and Missiles*, Annapolis, Md.: Nautical and Aviation Publishing Company of America, 1979, pp. 7–38.

[22] Thirty-five squadrons were equipped with KC-97s; four had yet to transition to the new aircraft and were still flying the obsolete KB-29. See Polmar, 1979, p. 38.

[23] Polmar, 1979, pp. 13, 38, 48.

[24] The contemporary force sizing metric for the tanker fleet is the number of aircraft. For example, *QDR 2014* specifies 443 air refueling aircraft for the projected fiscal year 2019 force. See U.S. Department of Defense, *Quadrennial Defense Review 2014*, Washington, D.C.: Office of the Secretary of Defense, p. 40.

The Korean War

From a force presentation perspective, the Korean War was a time of transition from the group to the wing. The group, nevertheless, was more prominent and central to air operations, playing a role similar to that in World War II for most of the conflict. The strategic bombing campaign involved some large raids, involving multiple groups, but these were a shadow of World War II 8th Air Force "maximum effort" missions involving an entire air division with hundreds of aircraft in subordinate wings, groups, and squadrons. In contrast, a Korean War strategic bombing "maximum effort" typically involved two to three groups flying every few days.[25] Most missions involved fewer than 100 aircraft, and bombers no longer flew in elaborate squadron, group, and wing formations. Robert Futrell describes the shift toward squadron-centric bombing:

> The operations planners finally specified three methods of attack for as many different sets of target conditions: squadrons in trail, bombing visually on squadron leaders; squadrons in trail, bombing by radar on squadron leaders; or a bomber stream of individual aircraft, bombing individually by radar.[26]

Wings were important as owners and operators of air bases, but in the early phase of the war, Headquarters Far East Air Force (FEAF) reallocated forces within the theater by moving squadrons and groups among existing wing locations rather than moving wings. Reinforcements from CONUS were deployed as active duty squadrons and groups, as well as Air Force Reserve wings. FEAF also experimented with reinforced wings, as well as various consolidated rear-area maintenance arrangements in Japan.[27]

After the Korean War, defense planning entered another turbulent period, with budget concerns once again competing with military requirements. With the Cold War in full force, the administration of President Dwight D. Eisenhower developed a strategy that sought to save money through reliance on airpower and nuclear weapons. The New Look strategy did not directly affect USAF force presentation constructs, though it did emphasize long-range bombers and nuclear weapons over fighters and conventional capabilities.

Perhaps the most interesting force presentation–related development of this period was TAC's invention of the Composite Air Strike Force (CASF). The CASF was in part the product of institutional competition between SAC and TAC. With SAC playing a dominant role in big war planning, TAC "became a hotbed of concern for the problems of limited war, in part to broaden TAC's mission orientation beyond a defensive strategy."[28] The CASF was

---

[25] Robert F. Futrell, *The United States Air Force in Korea: 1950–1953*, Washington, D.C.: Office of Air Force History, 1983, p. 187.

[26] Futrell, 1983, p. 190.

[27] Futrell, 1983, pp. 67–71, 640–643.

[28] Leighton, 2001, p. 667.

activated on July 8, 1955, with a standing headquarters of only 85 military personnel and "a small tactical air force composed of a command element and of fighter, reconnaissance, tanker, troop carrier, and communications support units" available to it as contingencies required.[29] The CASF could deploy a squadron-size element to the Middle East in as little as 16 hours, and up to two composite wings within two days. Its primary purpose was deterrence and crisis management rather than warfighting, but it had sufficient capabilities to conduct initial air operations against lesser powers. During its 18-year existence, the CASF was deployed a number of times, including in 1958, to overlapping crises in Lebanon and Taiwan. Although disbanded in 1973, the CASF represented a serious effort to improve USAF rapid response capabilities and could be considered a precursor of the AEF that would follow it roughly forty years later.[30]

The Vietnam War

USAF involvement in the Vietnam War began small with the temporary deployment in November 1961 of a detachment of 12 aircraft and 140 men from the 4400th Combat Crew Training Squadron (CCTS), Eglin Air Force Base, Florida.[31] Upon arrival at Bien Hoa Air Base, Vietnam, the unit was designated Detachment 2A, 4400th CCTS with the code name Farm Gate.[32] To support Farm Gate and other units at Bien Hoa, the USAF created Detachment 9, 13th Air Force.[33] Also in November, the 13th Air Force activated a forward element of the 2nd Air Division at Tan Son Nhut Air Base (with detachments at Da Nang and Nha Trang) to oversee the growing USAF presence.[34]

Over the coming months it became clear that larger and more permanent organizations were required to support the U.S. effort. Although the USAF had previously moved to the wing/base model as the standard, the Vietnam mission had not yet risen to that scale. Consequently, combat air groups were created and used in Vietnam between 1962 and 1965. The 1st Commando Group (TAC) was created in April 1962 to house and support the Farm

---

[29] Richard G. Davis, *Anatomy of a Reform: The Expeditionary Aerospace Force*, Washington, D.C.: Air Force History and Museums Program, 2003, p. 4.

[30] Davis, 2003, pp. 4–7.

[31] This detachment represented roughly 40 percent of the squadron's personnel. This air commando unit (the 4400th CCTS, nicknamed Jungle Jim) was created a few months earlier (in April) at Eglin Air Force Base, Florida, for the dual purpose of training partner nation air forces in counterinsurgency and conducting combat operations, both with specially modified vintage aircraft, including the B-26, C-47, and T-28. See Robert F. Futrell, *The United States Air Force in Southeast Asia: The Advisory Years to 1965*, Washington, D.C.: Office of Air Force History, 1981, pp. 79, 271.

[32] For more details on Farm Gate, see Futrell, 1981, pp. 79–84.

[33] Futrell, 1981, pp. 81, 271.

[34] The 2nd Air Division was expanded and redesignated the 7th Air Force in 1966. William W. Momyer, *Air Power in Three Wars*, Washington, D.C.: Office of Air Force History, 1985, pp. 67, 274.

Gate detachment. The 33rd and 34th Tactical Groups were activated in 1962, each serving roughly two years as independent air groups. The 33rd became a Combat Support Group in 1965 and the 34th was redesignated the 6251 Tactical Fighter Wing in 1965.

By early 1966 all USAF flying units were either squadrons or wings. USAF force presentation in Vietnam largely followed the historic model in which most squadrons were permanent units deployed to the conflict whereas most wings were activated in theater for conflict-specific needs. Thus, of the 64 air combat squadrons that served in Vietnam or Thailand, 70 percent deployed from outside the theater (45 squadrons) but only 30 percent (19 squadrons) were activated in theater. The reverse was true for wings. Of the 20 tactical fighter wings or air commando wings that served in Vietnam or Thailand, only 35 percent (seven wings) transferred from outside the theater; 65 percent (13 wings) were activated in theater.[35]

For the air war over North Vietnam—which often involved large numbers of combat and support aircraft—the wing played a central role in mission planning and execution.[36] Although the wing normally planned and executed missions based on orders received from higher commands,[37] in one famous 1967 operation, the 8th Tactical Fighter Wing (TFW) proposed and led a multiwing 7th Air Force operation designed to counter the growing North Vietnamese fighter threat. Operation Bolo, the brainchild of 8th TFW wing commander Col Robin Olds and Capt John B. Stone, the wing tactics officer, was a ruse designed to reduce MiG-21 attacks on USAF strike missions—which were causing heavy losses among bomb-laden F-105 fighters.

Colonel Olds won the enthusiastic support of 7th Air Force commander Gen William Momyer. Working with 7th Air Force leadership and staff, the 8th TFW crafted a mission that perfectly mimicked F-105s.[38] Led by Colonel Olds, the task force included elements of the 8th,

---

[35] Fighter, bomber, tactical air support, tactical reconnaissance, and tactical electronic warfare squadrons and air commando, tactical fighter, tactical reconnaissance and fighter wings operating from bases in Thailand or Vietnam are counted in this analysis. Calculations are based on data compiled from Appendix 1, "United States Order of Battle," in Chris Hobson, *Vietnam Air Losses: United States Air Force, Navy and Marine Corps Fixed-Wing Aircraft Losses in Southeast Asia, 1961–1973*, Hinckley, England: Midland Publishing, 2001, pp. 250–257.

[36] For more on mission planning at the wing and squadron levels during the air war against North Vietnam, see Jack Broughton, *Thud Ridge*, New York: J. B. Lippincott Company, 1969, pp. 248–249.

[37] Although wing staff played important roles in mission planning and execution, it should be noted that higher headquarters, from the 7th Air Force to the DOD, Joint Chiefs of Staff, National Security Council, and White House, all played large and controversial roles in target selection and rules of engagement. See Mark Clodfelter, *The Limits of Air Power: The American Bombing of North Vietnam*, New York: The Free Press, 1989; Robert A. Pape, *Bombing to Win: Air Power and Coercion in War*, Ithaca, N.Y.: Cornell University Press, 1996, pp. 174–210; and Momyer, 1985.

[38] Command and control arrangements during the Vietnam War were complex. Although the 8th TFW was under the command of the 13th Air Force (headquartered in the Philippines), the 7th Air Force at Tan Son Nhut Air Base had operational control of all 13th Air Force units in Thailand. See Carl Berger, ed., *The United States Air Force in Southeast Asia: 1961–1973*, Washington, D.C.: Office of Air Force History, 1977, p. 81.

355th, 366th, and 388th TFWs. The 7th Air Force provided supporting EB-66 EW aircraft, KC-135 tankers, EC-121 aircraft, search and rescue forces, and a C-130 airborne command post.[39]

F-4 phantoms in the strike package used "F-105 formations, attack tactics, and radio call signs" to convince North Vietnamese air defense controllers that a large package of F-105 strike aircraft was heading for Hanoi.[40] A large force of MiG-21s was launched to intercept but found, instead of F-105s, the much more capable F-4s fully loaded for air combat. The result was between seven and nine MiGs shot down and no USAF losses.[41]

In sum, Vietnam followed the historic practice in which tactical execution was largely a squadron responsibility, with wings playing critical roles in mission planning, maintenance, security, medical, communications, and other support functions and numbered air forces orchestrating larger packages as well as generating daily operations orders.

From Vietnam to the Fall of the Berlin Wall, 1975–1989

After Saigon fell to the North Vietnamese in April 1975, the USAF returned its full attention to the Soviet and Warsaw Pact military challenge in Central Europe. Force planning was driven by the needs of this theater between 1975 and the fall of the Berlin Wall in 1989. Although the Soviet Union was not dissolved until 1991, for our purposes it is more helpful to end this discussion in 1989. The creation of the objective wing in 1990 and Operations Desert Shield (1990) and Desert Storm (1991) are more closely tied to post–Cold War developments and will be discussed in the final historical section of this chapter.

The squadron continued as the USAF force presentation construct for deployment. The DOD's *Annual Report* for fiscal year 1985 uses this construct as well, observing, "The Air Force prepositions equipment in Europe to *speed the delivery of tactical fighter squadrons* and the minimum support they need to begin fighting."[42] Although units deployed as squadrons, wings continued to provide most support functions. Under the wing/base model, wings also operated bases at home and abroad. In cases where flying units were not permanently based, air base wings provided support functions for visiting units.

---

[39] Robin Olds with Cristina Olds and Ed Rasimus, *Fighter Pilot: The Memoirs of Legendary Ace Robin Olds*, New York: St. Martin's Press, 2010, pp. 269, 271–282.

[40] Benjamin S. Lambeth, *The Transformation of American Air Power*, Ithaca, N.Y.: Cornell University Press, 2000, p. 45.

[41] Brian D. Laslie, *The Air Force Way of War: U.S. Tactics and Training After Vietnam*, Lexington: University Press of Kentucky, 2015, pp. 15–16.

[42] Caspar W. Weinberger, *Annual Report to the Congress, Fiscal Year 1985*, Washington, D.C.: Office of the Secretary of Defense, February 1, 1984, p. 183, emphasis added.

A 1987 *Airpower Journal* article captured the central role of the wing as a Cold War fighting organization:

> The wing commander is fully responsible for the tactical employment of his aircraft to achieve tactical, operational or strategic aims. At the same time, he occupies a position at the bottom fringe of the operational level of war . . . . At the tactical level, the wing commander must ensure that he has a secure base from which to fight and that he is getting his best from the available logistic support in executing his tasking. He must orchestrate the execution of the wing's tasking and he must be a leader and commander for both his flying and support forces . . . . At the operational level of war, the wing commander ensures that his wing is always prepared for shifts in the air campaign . . . . The wing commander needs to understand his own commander's concept for conducting the war . . . [and] must advise his commander as to the ability of his wing to support each of the ongoing or expected air campaigns.[43]

The wing remained the most common force-sizing metric for the fighter force, though squadrons and even groups were used to size other force elements.[44] Between 1975 and 1989, eight DOD annual reports used the wing as the fighter force metric, three reports used the squadron to size the fighter force, and three reports measured fighter force size in both squadrons and wings. Squadrons were also used at times to size bomber units and forces within regions, as seen in the DOD's *Annual Report* for fiscal year 1985, which describes the USAF combat force as consisting of 37 tactical fighter wings but 7 bomber squadrons. The DOD report also used squadron numbers to describe the size of a variety of smaller forces such as electronic warfare, tactical reconnaissance, and Special Operations Forces units. Finally, it used the squadron metric for U.S. Air Forces in Europe, noting that the United States maintained "28 Air Force Squadrons" in Europe; the same report also listed "7 Tactical Fighter Wings" and "2 Strategic Bomber Squadrons" among the forces available to U.S. Central Command (CENTCOM).[45]

Operations Desert Shield and Desert Storm

With the end of the Cold War the USAF went through the first major reorganization since the 1950s shift to the wing-centric force. The most common Cold War wing and base organization was the tri-deputy form, with deputy commanders for operations, maintenance, and resources under a wing commander. The squadron remained the bedrock of the USAF. As the USAF's *Gulf War Air Power Survey* (*GWAPS*) from 1993 notes, "The squadron was the

---

[43] Clifford R. Krieger, "Fighting the Air War: A Wing Commander's Perspective," *Airpower Journal*, Vol. I, No. 1, Summer 1987.

[44] This is based on a RAND review of 33 DOD *Annual Reports* published between 1969 and 2005 and available from the OSD's Historical Office; see Office of the Secretary of Defense, Historical Office, *Secretary of Defense Annual Reviews*, website; note that reports are not available for every year.

[45] Weinberger, 1984, pp. 204, 212.

deployable unit in the Air Force. War plans tasked organizations by squadron not by wing, and support revolved around providing for, employing, and sustaining squadrons."[46]

Declining force structure and budgets led USAF leaders to emphasize "elimination of unnecessary layers of authority, decentralization of decision-making, and consolidation of functions." In 1990, the USAF created the objective wing, a structure similar to the 1947 wing-base plan. Like the 1947 plan, the new wings had four groups: operations (formerly combat), logistics, support, and medical. Most squadrons were aligned under one of these four groups. The reorganization sought to retain "a high combat capability while increasing the operational flexibility of the much-reduced force."[47]

Also in August 1990, Iraq invaded Kuwait, triggering a massive U.S. military response in Operation Desert Shield. As was its practice, the USAF deployed at the squadron level. Historian Dick Hallion describes the first movement of fighters from the 1st Tactical Fighter Wing:

> On August 7, at 5:25 PM East Coast time, twenty-four F-15C air superiority fighters of TAC's 71st Tactical Fighter Squadron thundered aloft from Langley AFB in Tidewater Virginia. They landed at Dhahran on the afternoon of August 8 (Saudi time), having completed a 15-hour, 8,000 mile, nonstop journey from the United States involving up to twelve in-flight refuelings.[48]

Two days later, on August 9, a second squadron (the 27th TFS) departed Langley Air Force Base for Dhahran. This was fairly typical for the 13 TFWs that deployed two or more squadrons. Seven of the 13 TFWs closed in five or fewer days, but in no case did a TFW deploy multiple squadrons on the same day (see Table 3.1). Six TFWs took over 100 days to fully close—further evidence that the squadron was the key level for unit deployments rather than the wing.

The USAF created provisional air divisions, wings and, to a lesser degree, squadrons to command and control these forces.[49] The *GWAPS* records 29 wing-level organizations participating in the Gulf War. Of these, 28 were provisional units. In contrast, only 38 of the 101 subordinate units (mainly squadrons) were provisional.[50] With regard to command and

---

[46] U.S. Air Force, *Gulf War Air Power Survey:* Vol. III, *Logistics and Support*, Washington, D.C.: Department of the Air Force, 1993b, p. 49.

[47] U.S. Air Force, *A Guide*, undated.

[48] Richard P. Hallion, *Storm over Iraq: Air Power and the Gulf War*, Washington, D.C.: Smithsonian Institution Press, 1992, p. 136.

[49] The *GWAPS* reports that Gen Chuck Horner, the JFACC, created two provisional air divisions on December 5, 1990. The 14th Air Division was given "operational control of assigned tactical fighter wings," while the 15th Air Division controlled electronic warfare, reconnaissance, command and control, and other units. See U.S. Air Force, *Gulf War Air Power Survey:* Vol. I, *Planning and Command and Control*, Washington, D.C.: Department of the Air Force, 1993a, p. 70.

[50] Calculations are based on data presented in U.S. Air Force, *Gulf War Air Power Survey:* Vol. V, *A Statistical Compendium and Chronology*, Washington, D.C.: Department of the Air Force, 1993c, Table 4, pp. 22–25.

Table 3.1. USAF Fighter Wing Deployments During Operation Desert Shield

Wing	Squadrons Deployed	First Deployed	Last Deployed	Days To Close
20th TFW	3	1-Aug-90	17-Jan-91	170
1st TFW	2	8-Aug--90	9-Aug-90	2
4th TFW	4	10-Aug-90	3-Jan-91	147
336th TFW	2	10-Aug-90	11-Aug-90	2
354th TFW	2	18-Aug-90	20-Aug-90	3
37th TFW	2	21-Aug-90	4-Dec-90	106
48th TFW	4	25-Aug-90	11-Dec-90	109
33th TFW	3	29-Aug-90	2-Sep-90	5
401st TFW	2	29-Aug-90	18-Jan-90	153
388th TFW	2	30-Aug-90	1-Sep-90	3
23rd TFW	2	31-Aug-90	2-Sep-90	3
52nd TFW	4	5-Sep-90	17-Jan-91	135
36th TFW	2	16-Dec-90	20-Dec-90	5

SOURCE: RAND analysis of data compiled from U.S. Air Force, *Gulf War Air Power Survey:* Vol. V, *A Statistical Compendium and Chronology*, Washington, D.C.: Department of the Air Force, 1993c, Table 17, pp. 58–64.

control, Operation Desert Storm saw the first use of the Joint Force air component commander (JFACC) and Air Tasking Order (ATO) to direct (most) air operations. Gen Charles Horner, the JFACC, delegated control of the ATO to Brig Gen Buster Glosson. Glosson held dual positions as chief air campaign planner and commander of the 14th Air Division (which controlled all USAF fighter and fighter-bomber wings).[51]

Wing commanders were very much combat leaders during the Gulf War. The *GWAPS* reports, "Two days after CENTCOM planners had issued the U.S.-Desert Storm plan, the chief CENTCOM air campaign planner, Brig Gen Buster Glosson, briefed his USAF wing commanders—the operators—on their role in the air campaign plan."[52] Similarly, in his

---

[51] Edward C. Mann III, *Thunder and Lightning: Desert Storm and the Airpower Debates*, Maxwell Air Force Base, Ala.: Air University Press, April 1994, p. 158.

[52] U.S. Air Force, 1993a, p. 10.

autobiography, Glosson emphasizes the importance of wing commanders as combat leaders and claims to have spoken to each one daily during the conflict.[53]

The daily ATO was transmitted to Wing Operations Centers, which, with their subordinate squadrons,[54] conducted detailed mission planning.[55] Thus, although the JFACC and ATO represented major (and controversial) innovations in the command and control of joint air operations,[56] from the perspective of the wing and squadron, orders (now in the form of the ATO) came from a higher echelon (in this case, the 14th Air Division), as they had in the Vietnam War and prior conflicts. The wing and squadron remained partners in the planning and execution of the orders (employing forces, in this report's terminology) as they had previously.

The Expeditionary Era: 1991 to the Present

The Gulf War ended in 1991 with the decisive defeat of the Iraqi military but with Saddam Hussein in power and determined to pursue policies that were unacceptable to U.S. leaders. As a result, the United States and its allies established two no-fly zones (NFZs) to inhibit Iraqi regime aggression against its own people. The northern NFZ was created to protect Iraqi Kurds as part of Operation Provide Comfort, which was succeeded by Operation Northern Watch (ONW). Between 1991 and 2003, U.S. and allied aircraft flew 75,000 ONW sorties out of Incirlik Air Base, Turkey. Operation Southern Watch (OSW) was established to protect the Shia population in southern Iraq from the regime. U.S. and allied aircraft operated out of bases in Saudi Arabia and Persian Gulf states as well as from aircraft carriers, flying a total of 150,000 sorties between 1991 and 2003.[57]

On average, ONW and OSW deployed 100–150 on a daily basis, with peaks over 250 aircraft during coercive strikes against Iraq (e.g., Operation Desert Strike in 1996 and Operation Desert Fox in 1998). Unlike Cold War deployments, which were smaller, short-lived, and to fully manned and equipped Main Operating Bases, ONW and OSW deployments were to host nation bases that lacked the support capabilities and infrastructure of USAF main operating bases; deployments were also much longer, lasting from 90 to 180 days. These and

---

[53] Buster Glosson, *War with Iraq: Critical Lessons*, Charlotte, N.C.: Glosson Family Foundation, 2003, pp. 90, 108–111, 167–168.

[54] Evidence that the squadrons continued to play a major role in mission planning was the U.S. Central Command Air Forces request for 20 Sentinel Byte work stations "to provide flying squadrons with automated intelligence support for mission planning." U.S. Air Force, 1993c, p. 40.

[55] U.S. Air Force, 1993a, p. 116.

[56] For more on interservice disagreements over the authorities of the JFACC during Operation Desert Storm, see Mann, 1994; and David E. Johnson, *Learning Large Lessons: The Evolving Roles of Ground Power and Air Power in the Post–Cold War Era*, Santa Monica, Calif.: RAND Corporation, MG-405, 2007, pp. 34–37.

[57] Karl P. Mueller, *Denying Flight: Strategic Options for Employing No-Fly Zones*, Santa Monica, Calif.: RAND Corporation, RR-423-AF, 2013, pp. 4–5.

other global demands (e.g., Balkan peacekeeping) led to a decline in overall USAF aircraft mission-capable rates between fiscal year 1991 and fiscal year 1998 and a recognition that the USAF needed a mechanism to better manage ongoing demands.[58]

In 1995 the USAF began to develop and use new expeditionary force constructs to meet these ongoing demands. After three years of experience and refinement, these concepts were codified in the Expeditionary Air Force (EAF) announced by USAF Chief of Staff Gen Michael Ryan and Secretary of the Air Force F. Whitten Peters on August 4, 1998. General Ryan described the EAF as "a new way of doing business that improved predictability and stability in personnel assignments and furnished the service with a powerful management tool to more efficiently align its assets with the needs of the warfighting Commanders in Chief."[59] The EAF divided USAF combat elements into ten AEFs that would be available in pairs over a 15-month rotational cycle.[60]

From a force presentation perspective, the AEF represented a significant change from past practice. The main innovation was the creation of the AEF concept and process to manage force rotations, a problem that the USAF had not faced before 1990. USAF doctrine is careful to delineate the AEF's purpose and boundaries:

> To address growth in diverse regional commitments, the Air Force established the air expeditionary force (AEF) concept as a means to provide Air Force forces and associated support on a rotational, and thus, a relatively more predictable basis. AEFs however only provide a source of readily trained operational and support forces. They do not provide for a commander (specifically, a commander, Air Force forces) or the necessary command and control mechanisms. Thus, AEFs by themselves are not discrete, employable entities. Forces sourced from AEFs should fall in on in-theater command structures, and link up with in-theater Air Force forces. Thus, while AEF forces may deploy, they stand up as part of an Air Force component (which may be in the form of an air expeditionary task force), not as their own warfighting entity. In short, the AEF is the mechanism for managing and scheduling forces for expeditionary use; the AETF is the Air Force warfighting organization normally attached to a JFC [Joint Forces Command].[61]

The AEF construct also created new organizations, including the AETF, the AEW, the Air Expeditionary Group (AEG), and the Air Expeditionary Squadron (AES).

USAF doctrine defines the AETF as "the generic title used when a provisional Air Force command echelon is needed between a NEAF [Numbered Expeditionary Air Force] and an air

---

[58] Davis, 2003, pp. 19–20.

[59] Gen Michael Ryan, quoted in Davis, 2003, p. iii.

[60] The rotational concept evolved considerably between the announcement in 1998 and rollout in October 1999. For a detailed history, see Davis, 2003.

[61] U.S. Air Force, "The Air Expeditionary Force," in *Annex 3-30: Command and Control*, Maxwell Air Force Base, Ala.: Curtis E. LeMay Center For Doctrine Development and Education, 2014b.

expeditionary wing." The AEW "is the generic title for a deployed wing or a wing slice within an AETF. AEWs normally carry the numerical designation of the wing providing the command element." Doctrine also allows for the creation of provisional AEWs—which have been used in Southwest Asia for many years. AEGs have more in common with World War II groups than home station groups in the modern USAF. "Unlike traditional 'home station' groups, which are functionally organized . . . expeditionary groups deployed independent of a wing structure should contain elements of all the functions to conduct semi-autonomous operations. An AEG is composed of a slice of the wing command element and some squadrons." Finally, the AES is "configured to deploy and employ in support of taskings" but "is not designed to conduct independent operations . . . . As such, an individual squadron or squadron element should not be presented by itself without provision for appropriate support and command elements." [62]

The AEF force management system, however, does not deploy complete squadrons or wings to create these expeditionary units. Rather, it uses UTCs to meet deployment requirements. For example, a "UTC for a 12-aircraft fighter package might include not only the aircraft and pilots, but also maintenance crews, spare parts and specialized equipment. Some UTCs consisted only of material; others, mostly of personnel."[63] Thus, expeditionary units are often made up of UTCs from multiple supporting wings and squadrons.

Although the UTC gives the USAF great flexibility and has allowed it to meet CCDR demands, it comes with a cost. Unit cohesion is lost when the expeditionary unit is composed of individuals from many units. At home, squadron members who did not deploy often have insufficient resources to meet other critical demands such as training, and they routinely work long hours covering undermanned squadron functions. Without calling out the AEF and UTC explicitly, in an exit interview as Chief of Staff, Gen Mark Welsh described the downside to deploying in this manner:

> So you don't pick up your Air Force wing or your group or your squadron and deploy it wholesale into a base in the Middle East. If you're a flying unit you go more intact than other units do . . . . When you deploy as a support function . . . you might have one person or two people from the same unit. It's not a big chunk of people from one unit. So the cohesion of your unit, kind of the sanctity of the squadron, the relationship between supervisors and supervisees, is split to support combat operations . . . we've got to figure out how we get back to more of a squadron-focused approach to everything . . . .[64]

The readiness and morale impacts of sustained high operational tempo are arguably exacerbated by the AEF/UTC model, but it is unlikely that, short of more force structure, the

---

[62] Quotes and AETF details from U.S. Air Force, "AETF Organization," in *Annex 3-30: Command and Control*, Maxwell Air Force Base, Ala.: Curtis E. LeMay Center For Doctrine Development and Education, 2014a.

[63] Davis, 2003, p. 71.

[64] Mark Welsh, "Exit Interview," conducted by *Air Force Magazine* staff on June 17, 2016, online.

USAF could meet the CCDR demands without the flexibility of the AEF/UTC approach. Some hope that the right force presentation construct (e.g., like the USN CSG) could protect the USAF from unsustainable demands. The historic evidence shows that this is not the case.[65] Rather, it suggests that force presentation concepts (including the CSG) are fairly weak reeds compared to the institutional and political force of a CCDR request for forces during an ongoing combat operation.

How have the five force presentation functions been addressed in this expeditionary period? As noted above, the squadron, AEF, and UTC are all used for the *deploying forces* function, depending on the requirement. *Employing forces* continues to require a team effort between either a squadron and wing, or the AEF equivalents in the AES/AEG or AES/AEW. The AEW typically has responsibility to *sustain effects*, but an independent AEG can as well. Between 1998 and 2016, the squadron and wing were the primary force-sizing metrics, but in 2003 and 2004 the DOD *Annual Report* sized the USAF as ten AEFs.[66] Finally, the *managing rotations* function has been met by the AEF system.

Historical Trends in Force Presentation

This chapter has offered a brief overview of the evolution of force presentation concepts in USAF history. Table 3.2 summarizes changes in force presentation constructs on the five dimensions salient for the USAF. The squadron is the one constant in USAF history, appearing at least once in four of the five areas. For major combat operations, USAF flying units have always deployed as squadrons. It is only in the post–Cold War era, with the requirement to support extended deployments abroad, that we see the UTC used as a construct for deploying individuals or capabilities.

The particular mission and the scale of conflict largely determine the appropriate level for force employment. For example, in World War I, pursuit aircraft often operated in flights or squadrons, but by the end of that war, bombers were operating in groups and wings. In World War II, force employment varied greatly across theater and mission, from highly dispersed fighter operations in the South Pacific to the massive 8th Air Force raids on Germany, where 15–20 groups would fly on a single mission. In conflicts since World War II, the USAF

---

[65] Analysis of twenty years of deployment data by RAND researcher Meagan Smith has found that CSGs routinely violated the USN preferred deployment cap of six months, typically due to combatant commander requests. See Meagan Smith, *Do Force Presentation Metrics Protect Service Readiness?* PAF-1P-376, December 13, 2016. Other unpublished work by RAND researchers Stacie Pettyjohn and Meagan Smith has found that in almost 900 joint operations conducted between 1946 and 2016, the services deployed according to their preferred construct between 62 percent and 81 percent of the time. This suggests that service force presentation constructs may have some influence, but do not significantly constrain combatant command request for forces.

[66] See Donald H. Rumsfeld, *Annual Report to the President and the Congress*, Washington, D.C.: Office of the Secretary of Defense, 2003, p. 170; and Donald H. Rumsfeld, *Annual Report to the President and the Congress*, Washington, D.C.: Office of the Secretary of Defense, 2004, pp. 135–136.

Table 3.2. USAF Force Presentation Constructs, 1913–2016

	Size Force	Deploy Forces	Employ Forces	Sustain Effects	Manage Rotations
1913-1917	Squadron	Squadron	Individual aircraft	Squadron	NA
WWI	Squadron	Squadron	Squadron, Group, Wing	Group	NA
Interwar	Group	Squadron	Squadron, Group, Wing	Group and Wing	NA
WWII	Group	Squadron	S, G, W & Air Division	Group	NA
Early Postwar	Group	Squadron	Squadron, Group	Group and Wing	NA
Korean War	Wing	Squadron	Squadron, Group	Wing	NA
Vietnam War	Squadron	Squadron	Squadron, Wing	Wing	NA
1976-1990	Squadron, Wing	Squadron	Squadron, Wing	Wing	NA
1998-2016	Squadron, Wing & AEF*	UTC, AES, Squadron	Squadron, Wing, AES, AEG, AEW	AEG, AEW or Wing	AEF

SOURCE: RAND analysis of historical materials; U.S. Department of Defense, *Annual Reports*, 1969–2005; and U.S. Department of Defense, *Quadrennial Defense Reviews*, 1997, 2001, 2006, 2010, 2014. For the period covered by the DOD *Annual Report*, the most frequent force-sizing metrics are listed, but they varied over time.
NOTE: * In the 2003 and 2004 reports, the USAF force-sizing metric was ten AEFs.

has employed force at the squadron and wing levels. Even in cases where smaller numbers of aircraft are required per mission, the wing makes critical contributions in mission planning.

A vast enterprise is required to sustain operational effects over time. The group was responsible for air base operation, aircraft maintenance, and other critical support functions for World War I and World War II. For the modern USAF, the wing (whether a permanent organization or an AEW) runs the air base and, through its subordinate groups, provides critical maintenance, logistical, security, medical, and other support.

The USAF has been sized using squadrons, groups, and wings as metrics. Two DOD *Annual Reports* (2003 and 2004) reported USAF size only as "10 Air Expeditionary Forces," but there is no evidence that any program or budget decisions were made using that construct. The AEF was not used for force sizing in any *QDR*. Rather, *QDR*s have used squadrons and wings to size the force. The most recent, *QDR 2014*, uses the squadron to size the fighter and bomber forces; the airlift force is presented in numbers of aircraft rather than units (common practice since the 1980s).

Finally, for most of its history, the USAF did not rotate forces abroad on temporary duty, so there was no *managing rotations* function. Since 1990, relatively small but enduring operations from expeditionary bases have become the norm, requiring a rotational base to support. The AEF is the only force presentation construct used by the USAF for this purpose.

This historical assessment suggests that the squadron is the most commonly used force presentation construct, at least as measured on these metrics. As Table 3.2 shows, the squadron was the sole construct 12 times and was paired with another construct in another 11 cases. Thus, the squadron was used in 23 of the 37 cases (~62 percent). In contrast, the group was the sole construct five times and paired in seven cases. The wing was the sole construct four times and paired in 11 cases.

4. Force Presentation in USAF Narratives

This chapter explores the role of force presentation constructs in USAF public narratives and in the writings of airpower theorists and historians. It documents the absence of force presentation in USAF narratives and major airpower writings and argues that airpower theory has historically followed a logic that, although sound, has been incomplete. The chapter ends with a proposed expansion to the logic that would allow the inclusion of force presentation.

Force Presentation in USAF Planning Documents and Airpower Theory

As was discussed in Chapter Three, the squadron is the most common force presentation construct across the functions assessed in this report. Long recognized as the pillar of the USAF, the squadron has received renewed attention over the last three years, primarily because of growing concerns that the combination of unending operational demands, smaller force structure, and reduced budgets is breaking the cohesion of the squadron.

Evidence of both the central role of the squadron and concerns about its health is found in the 2013 USAF vision statement. There is not a single mention of groups or wings in the vision, but an entire paragraph is devoted to the squadron. Also note the commitment to "reinvigorate" the squadron and the inclusion of "cohesive" and "ready" in the goals:

> The source of Air Force airpower is the fighting spirit of our Airmen, and squadrons are the fighting core of our Air Force. The evolving threats we face demand that our squadrons be highly capable, expeditionary teams who can successfully defend our Nation's interests in tomorrow's complex operating environments. We will reinvigorate squadrons and emphasize a unified chain of command, focused on mission success, and supported by centralized functional managers. Our squadrons will be the cohesive, ready, and agile fighting forces that the Air Force, the joint force commander, and the Nation require.[1]

The importance of the squadron was also emphasized in remarks that Gen Mark Welsh, the USAF chief of staff, made in June 2016 shortly before his retirement:

> And I think the importance of the squadron will be something the new Chief has got to keep focused on. The squadron is the foundational unit of our Air Force. It is the fighting unit of our Air Force. We have done a whole bunch of things that have not added to the cohesion and unity of squadrons in the way we've been deploying for the last 15 years. And squadrons are a way of addressing a whole lot of things—pride, morale, legacy, focus on heritage, the profession of arms, taking care of individuals and families . . . the root of where

[1] U.S. Air Force, *The World's Greatest Air Force—Powered by Airmen, Fueled by Innovation: A Vision for the United States Air Force*, Washington, D.C.: Department of the Air Force, 2013b, p. 2.

we're really good at that is in the squadron, so anything we can do to strengthen squadrons, we can damage a lot of other level organizations, but the squadrons have to work.[2]

Within a month of becoming USAF chief of staff, Gen David Goldfein published the first in a series of short papers laying out his thoughts on priority "focus areas" for the USAF; the paper identifies the squadron as the key echelon for the USAF and calls for its revitalization.[3] Although the squadron has a unique place in USAF organizational culture, senior airmen have long appreciated the importance of all the core tactical echelons (flights, squadrons, groups, and wings).

It is surprisingly, therefore, to find these echelons (and force presentation more broadly) to be largely absent from the primary USAF vision and planning documents, in the writings of prominent airpower theorists, and in major operational histories of the USAF in combat. Table 4.1 displays the number of times that force presentation concepts appear in the five major USAF planning and narrative documents.[4]

Table 4.1. Frequency with Which Force Presentation Concepts Appear in Major USAF Documents

Frequency with which word/phrase appears	GV/GR/GP (2013)	USAF Vision (2013)	USAF Strategy (2014)	SMP (2015)	AFFOC (2015)
Force presentation	0	0	0	1	0
Squadron	1	4	0	0	0
Wing	1	0	0	0	0
AEF	0	0	0	0	0
AETF	0	0	0	0	0
UTC	0	0	0	0	0
Expeditionary	0	1	0	3	2

NOTES: *GV, GR, GP* = U.S. Air Force, *Global Vigilance, Global Reach, Global Power for America*, Washington, D.C.: Department of the Air Force, 2013a. USAF Vision = U.S. Air Force, *The World's Greatest Air Force—Powered by Airmen, Fueled by Innovation: A Vision for the United States Air Force*, Washington, D.C.: Department of the Air Force, January 10, 2013b. USAF Strategy = U.S. Air Force, *America's Air Force: A Call to the Future*, Washington, D.C.: Department of the Air Force, July 2014d. *SMP* = U.S. Air Force, *Strategic Master Plan Executive Summary*, Washington, D.C.: Department of the Air Force, 2014d. AFFOC = Air Force Future Operating Concept.

---

[2] Welsh, 2016, online.

[3] David Goldfein, *CSAF Focus Area: The Beating Heart of the Air Force . . . Squadrons!* Washington, D.C.: Department of the Air Force, August 2016.

[4] All five documents are available online as searchable PDF files. The word count was done using the PDF search function.

None of these major planning documents discusses the concept of force presentation directly. Tactical echelons and organizational constructs are almost entirely absent as well; with the exception of the squadron (called out in the USAF vision statement as "the fighting core of our Air Force"), they are ignored. The complete absence of the AEF in these documents is noteworthy. Many in the USAF would point to it as the USAF force presentation construct, yet neither "AEF," "AETF," or "UTC" appear in these documents.

Force presentation is also absent in prominent airpower writings.[5] Table 4.2 displays the results of word searches in seven well-known works by airpower theorists.[6] Modern concepts like the AEF are, of course, absent from early works, but they make few appearances in modern works. The terms *squadron* and *wing* appear more frequently, but these are primarily descriptive, as in "Captain Lawson led his squadron" (Mitchell) or "as one fighter wing

Table 4.2. Frequency with Which Force Presentation Concepts Appear in Selected Airpower Writings

Concept/ Representative Pubs	Mitchell (1925)	de Seversky (1942)	Warden (1995)	Meilinger (1996)	Lambeth (2000)	Deptula (2001)	Mueller (2010)
Force presentation	0	0	0	0	0	0	0
Squadron	20	8	0	1	17	0	0
Wing	0	Unk	0	0	14	0	0
AEF	NA	NA	0	0	7	0	0
AETF	NA	NA	0	0	0	0	0
UTC	NA	NA	0	0	0	0	0
Expeditionary	0	11	0	0	9	0	0

NOTES: The works referred to in this table are, William Mitchell, *Winged Defense: The Development and Possibilities of Modern Air Power—Economic and Military*, Tuscaloosa: University of Alabama Press, 2009; Alexander P. de Seversky, *Victory Through Air Power*, New York: Simon and Schuster, 1942; John Warden, "The Enemy as a System," *Airpower Journal*, Vol. IX, No. 1, Spring 1995, pp. 40–55; Phillip S. Meilinger, "Ten Propositions Regarding Air Power," *Airpower Journal*, Vol. X, No. 1, Spring 1996, pp. 1–18; Benjamin S. Lambeth, *The Transformation of American Air Power*, Ithaca, N.Y.: Cornell University Press, 2000; David Deptula, *Effects-Based Operations: Change in the Nature of Warfare*, Arlington, Va.: Aerospace Education Foundation, 2001; Karl P. Mueller, "Air Power," in Robert A. Denemark, ed., *The International Studies Encyclopedia:* Vol. I, Oxford: Wiley-Blackwell, 2010, pp. 47–65.

[5] For a comprehensive treatment of the evolution of airpower theory, see Phillip S. Meilinger, ed., *The Paths of Heaven: The Evolution of Airpower Theory*, Maxwell Air Force Base, Ala.: Air University Press, 1997.

[6] The Deptula, Meilinger, Mueller, and Warden writings were available as searchable PDF files. The Lambeth, Mitchell, and de Seversky books were available in hard copy and as searchable files at the Google Books website.

commander reported" (Lambeth).[7] Although the details of force presentation matter to the USAF, it has not been and is not central to airpower narratives. Why is this the case?

American airpower narratives written by Mitchell, de Seversky, Warden, and Deptula represent the most public presentations of airpower theory from the interwar period to today. These narratives do not necessarily represent official USAF thinking, which tends to be cautious and less crisp in its arguments about airpower. They do, however, contain themes that have resonated with many airmen over the last 100 years.[8] These narratives follow a framework that begins with desired effects and ends with required forces. Although the specifics of the narratives have varied over the intervening years, the basic logic has endured. For a given set of effects, specific missions and targets are identified, and force requirements flow naturally from these demands. Figure 4.1 illustrates this logic using the dominant airpower concept of the interwar years as presented by Mitchell in his 1925 book *Winged Defense*.

Figure 4.1. Billy Mitchell's Airpower Narrative

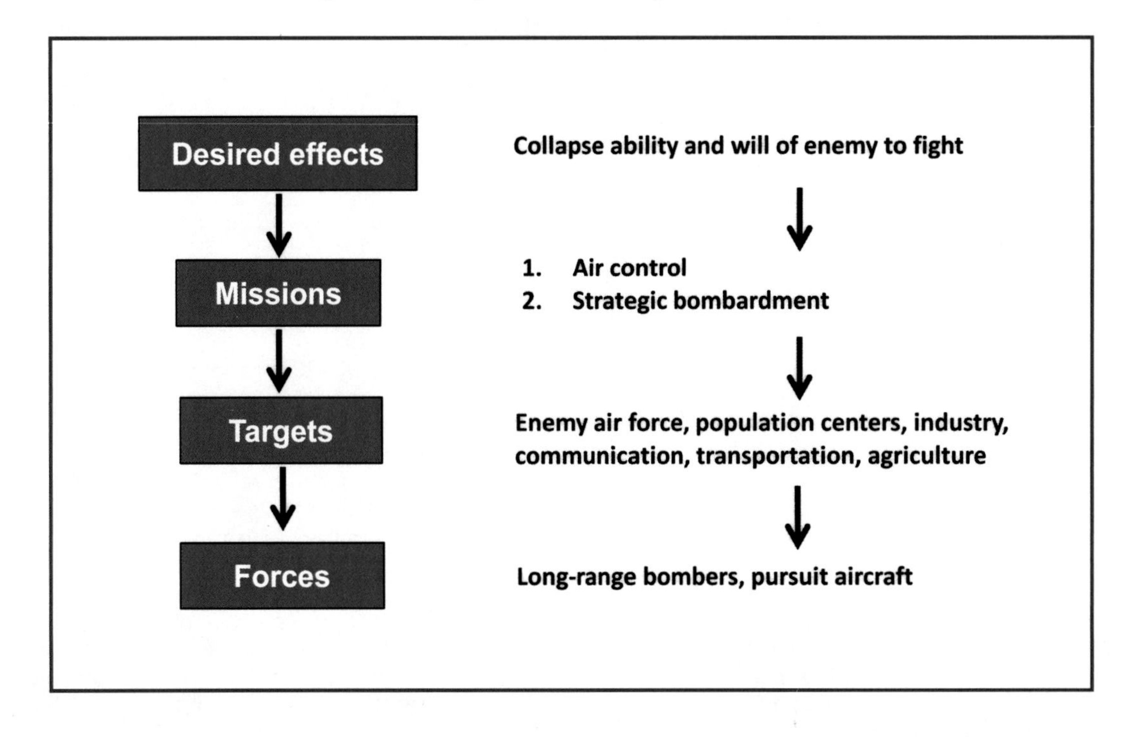

---

[7] William Mitchell, *Winged Defense: The Development and Possibilities of Modern Air Power—Economic and Military*, Tuscaloosa: University of Alabama Press, 2009, p. 65; Benjamin S. Lambeth, *The Transformation of American Air Power*, Ithaca, N.Y.: Cornell University Press, 2000, p. 38.

[8] For a more complete discussion of trends in airpower narratives over time, see Paula G. Thornhill, *"Over Not Through": The Search for a Strong, Unified Culture for America's Airmen*, Santa Monica, Calif.: RAND Corporation, OP-386-AF, 2012; and Vick, 2015.

This narrative framework is logical and compelling to describe the rationale for aircraft and other force structure needs, but it does not tell the whole story. In particular, this approach leaves important policy questions about force presentation and basing unanswered: (1) How does the USAF present forces? (2) What is the role of the air base? (3) Why are forward bases necessary?

The USAF does not routinely present forces in large formations, as the other services do via the Army BCT, the MAGTF, or the CSG. It can deploy multiple wings for major conflicts, but more typically offers its forces to the CCDR as squadron or smaller packages of capabilities (via the UTC). Although the USAF uses the AEF construct to generate and manage deployments, the AEF is not particularly helpful for the USAF narrative, in part because some important missions are conducted by forces that rarely or never deploy. A more accessible narrative would simply note that the USAF presents forces in a uniquely flexible and scalable manner ranging from individual aircraft, space platforms, or cybermission teams to hundreds of platforms and thousands of airmen as the contingency requires. This flexibility is recognized by combatant commanders who routinely ask for airpower in increments as small as individual aircraft. This, admittedly, is a mixed blessing given the stresses that meeting small force requests have placed on squadrons, but it is a demonstration of the great agility of the USAF.

The USAF is able to generate forces with such agility for a variety of reasons, but one is routinely neglected: the USAF global network of fixed facilities. Unlike the Navy (which generates combat power from the fleet at sea), the Army (which generates combat power from maneuver units in the field), or the Marines (which do both), the USAF creates operational effects from *bases*. Gen Henry H. Arnold captured the importance of basing in his classic 1941 observation that "air bases are a determining factor in the success of air operations. The two-legged stool of men and planes would topple over without this equally important third leg."[9]

These bases can support a wide range of force types, sizes, and missions. From these locations the USAF integrates effects across domains via the Air Operations Center, conducts cybermissions, launches and controls space systems, and generates aircraft sorties. In short, the USAF fights from its bases—on U.S. soil and abroad. The forward air base, in particular, is essential for many combat missions, and it remains the most visible manifestation of airpower. Although U.S. basing strategy increasingly emphasizes access to dispersal bases rather than building new main operating bases, an enduring USAF presence in key regions is foundational to deter aggression, reassure partner nations, and—if necessary—fight wars.

Figure 4.2 illustrates an expanded USAF narrative that incorporates these additional considerations. It begins with desired effects. In 2018 these are broader than those effects of

---

[9] Henry H. Arnold, "The Air Forces and Military Engineers," *The Military Engineer*, Vol. XXXIII, No. 194, December 1941, p. 548.

Figure 4.2. Incorporating Force Presentation into the USAF Narrative

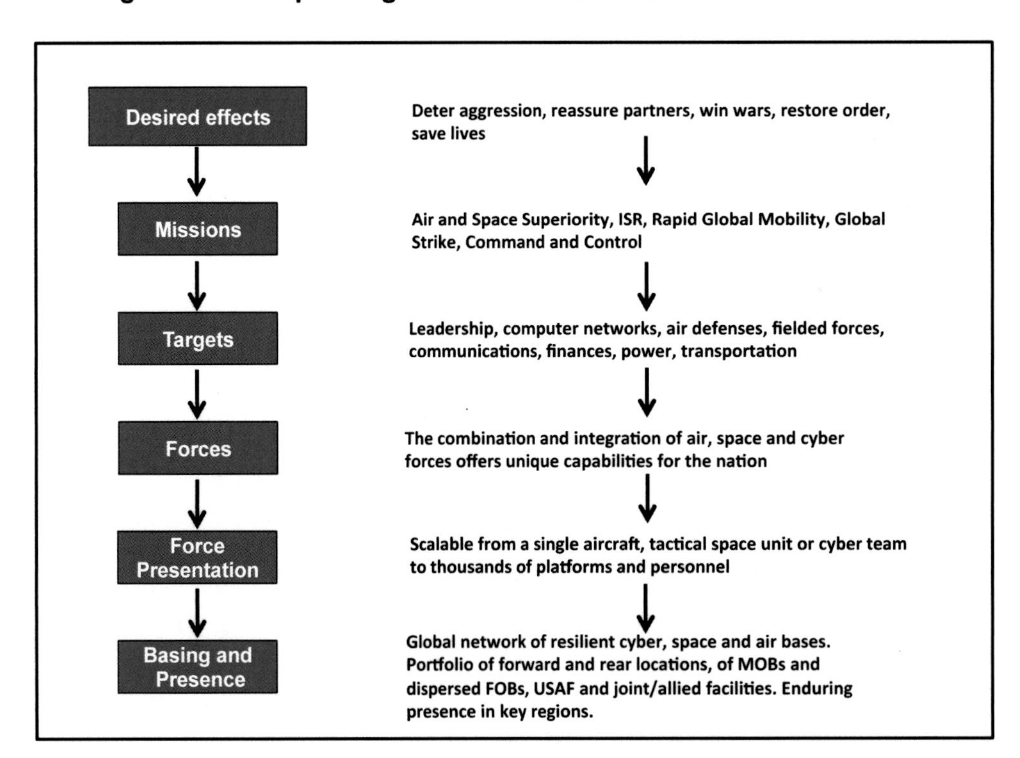

concern to airmen writing during the interwar years. These effects include deterring aggression, reassuring partners, restoring order, saving lives, and—when necessary—winning wars. Missions flow from these desired effects. Although missions can be expressed in a variety of ways, the USAF five mission areas capture the essence of its contributions: air and space superiority; intelligence, surveillance, and reconnaissance (ISR); rapid global mobility; global strike; and command and control.[10] Targets for USAF operations (both kinetic and non-kinetic) include enemy leadership, computer networks, air defenses, fielded forces, communications, finances, power, and transportation. These desired effects, missions, and targets generate force requirements. Although the USAF must defend specific programs such as the F-35, its high-level narrative needs to be broader. For example, it might emphasize how the combination and integration of air, space, and cyber forces offers unique capabilities for the nation. With that high-level narrative in place, it is much easier to then argue for the specific programs. In the case of the F-35, its unique ability to collect and move information in a combat network ties nicely to this high-level narrative by emphasizing the integration of forces across domains. So far we have simply updated the earlier effects-to-missions-to-targets-to-forces logic. We now add force presentation.

[10] The importance of the wing and squadron structure varies across these mission areas. It is most central to flying units in the CAF.

As noted earlier, an accessible external narrative needs to focus on the essence of USAF capabilities rather than the details of internal processes. Thus, rather than talking about the AEF or some other internal construct, the narrative should speak to the agility and scalability of USAF capabilities across platforms and domains. For example, using language more appropriate for the general public, it might note that the USAF has the unique ability to conduct effective military missions with very small forces (e.g., a single aircraft, satellite constellation, or cyberteam) or rapidly deploy much larger forces involving thousands of platforms and personnel.[11] For specialized audiences such as the Office of the Secretary of Defense (OSD) and Congress, the number of AEFs, squadrons, or wings will be of interest, but the primary message is about agility and scalability not specific echelons.

Just as there is no single all-purpose force presentation construct, there also is no one narrative that will address every USAF need or speak to every audience. The narrative framework presented here is designed to incorporate force presentation, basing, and presence into a high-level airpower narrative whose logic is accessible to a wide audience. Narrower narratives—for example, regarding the central role of the squadron in USAF culture and operations—may be called for to build consensus within the USAF for necessary reforms, as well as to carry that message to more expert audiences in OSD, the other services, the joint staff, combatant commands, and Congress.

Public Perceptions of Force Presentation

The broader public primarily experiences force presentation with regard to the *deploying forces* function. Reports of force deployments occur at four levels: the institutional, organizational, platform, and individual. An example of the institutional level is found in a *CNN Online* headline reporting "U.S. Army Sending Armored Convoy 1,100 Miles Through Europe." The organizational level is found in reports such as the *Navy Times* article titled "The U.S. Just Sent a Carrier Strike Group to Confront China." The platform level is most relevant for the USAF and captured by another *CNN Online* headline, "U.S. Sends F-22 Warplanes to Romania." Finally, force presentation is also seen at the individual level, as in the *Marine Corps Times* headline "2,300 California-Based Marines Just Deployed to the Middle East."[12] This section briefly discusses public interest in the organizational and platform levels as measured by internet searches as reported by Google Trends. Google Trends is increasingly

---

[11] For a crisp rethinking and updating of core tenets of airpower (and, in particular, how precision has redefined the concept of mass), see Phillip S. Meilinger, "Ten Propositions Regarding Air Power," *Airpower Journal*, Vol. X, No. 1, Spring 1996, pp. 1–18.

[12] See Brad Lendon, "U.S. Army Sending Armored Convoy 1,100 Miles Through Europe," *CNN Online*, March 14, 2015; David B. Larter, "The U.S. Just Sent a Carrier Strike Group to Confront China," *Navy Times*, March 3, 2016; Clarissa Ward, "U.S. Sends F-22 Warplanes to Romania," *CNN Online*, April 26, 2016; and Matthew L. Schehl, "2,300 California-Based Marines Just Deployed to the Middle East," *Marine Corps Times*, April 22, 2016.

used as a research tool, offering insights regarding the level of public interest in a topic. That said, it can be tricky to use and techniques to mine it for social insights are still in development. For these reasons, the results presented here should be viewed as hypotheses for further investigation rather than as definitive conclusions.[13]

Figure 4.3 is a screenshot of a Google Trends display for four organizational-level presentation constructs: "Brigade Combat Team," "Carrier Strike Group," "Marine Expeditionary Unit," and "Air Expeditionary Wing."[14] The Brigade Combat Team (BCT)

Figure 4.3. Relative Interest in Organizational-Level Force Presentation

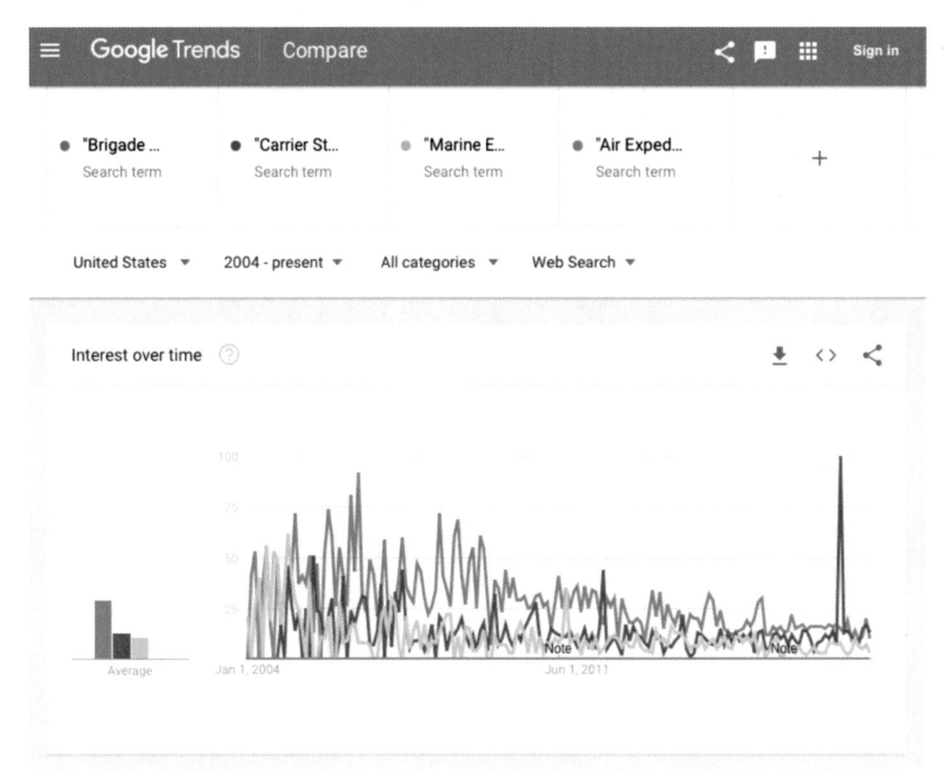

SOURCE: Google Trends search, December 19, 2017.

---

[13] Thanks to Col Robert J. Levin, USAF, for his constructive suggestions on this section of the report.

[14] Google Trends is a tool that shows the volume of Google online searches over time. Up to five search terms or phrases can be compared. It is used by marketers to track interest in products, ideas, and phrases. It also has been used by researchers to, for example, track the spread of influenza by region based on search volume. Google Trends outputs are based on a sample of searches and are normalized. Search term choice is critical and can lead to wide variations in outputs. For this study, the author first conducted Google searches using variations such as "BCT," "Brigade," and "Brigade Combat Team" to determine which terms produced the most accurate results. Abbreviations are particularly risky and were avoided in most cases. For example, the top hits for "BCT" were (1) Business Cards Tomorrow; (2) a Wikipedia article on all the different meanings of that abbreviation; (3) Beach Cities Transit in Redondo Beach, California; and (4) BCT Entertainment. Thus, "BCT" was not used. For more on Google Trends benefits and limitations in research, see Jonathan Mellon, "Where and When Can We Use Google Trends to Measure Issue Salience?" *Political Science and Politics*, Vol. XLVI, No. 2, April 2013, pp. 280–290; and Galen Stocking and Katerina Eva Matsa, "Using Google Trends Data for Research? Here Are 6 Questions to Ask," Pew Research Center, December 19, 2017.

is the organizational construct of greatest interest to the public as measured by internet searches, reaching peak interest in 2006 during the height of fighting in Iraq. Searches for "Carrier Strike Groups" also reflect events in the news, spiking in April 2017 when a CSG was deployed in response to a North Korean missile test. In contrast, "Air Expeditionary Wing" was searched for the least often.

Figure 4.4 compares force presentation at the platform level. Admittedly this is a somewhat less natural comparison than the organizational one, but it is still instructive. The search terms were "F-22," "aircraft carrier," "MRAP," and "MV-22," representing the most frequently searched for platforms associated with each service. The F-22 and aircraft carrier compete for first place, with the aircraft carrier receiving the greatest volume of hits. This comparison is not, however, entirely fair because "aircraft carrier" can refer to any such platform in the USN or other navies. In contrast, "F-22" is a specific reference to one aircraft type that is only in service with the USAF. Unfortunately, alternatives like "CVN," or "Nimitz-class carrier" are terms that few outside the Navy or defense community would use. Thus, we are left with a somewhat ambiguous result.

Of particular interest for those seeking to better integrate force presentation in the USAF narrative is the comparison in Figure 4.5 between a USAF platform (the F-22) and all service

Figure 4.4. Relative Interest in Platform-Level Force Presentation

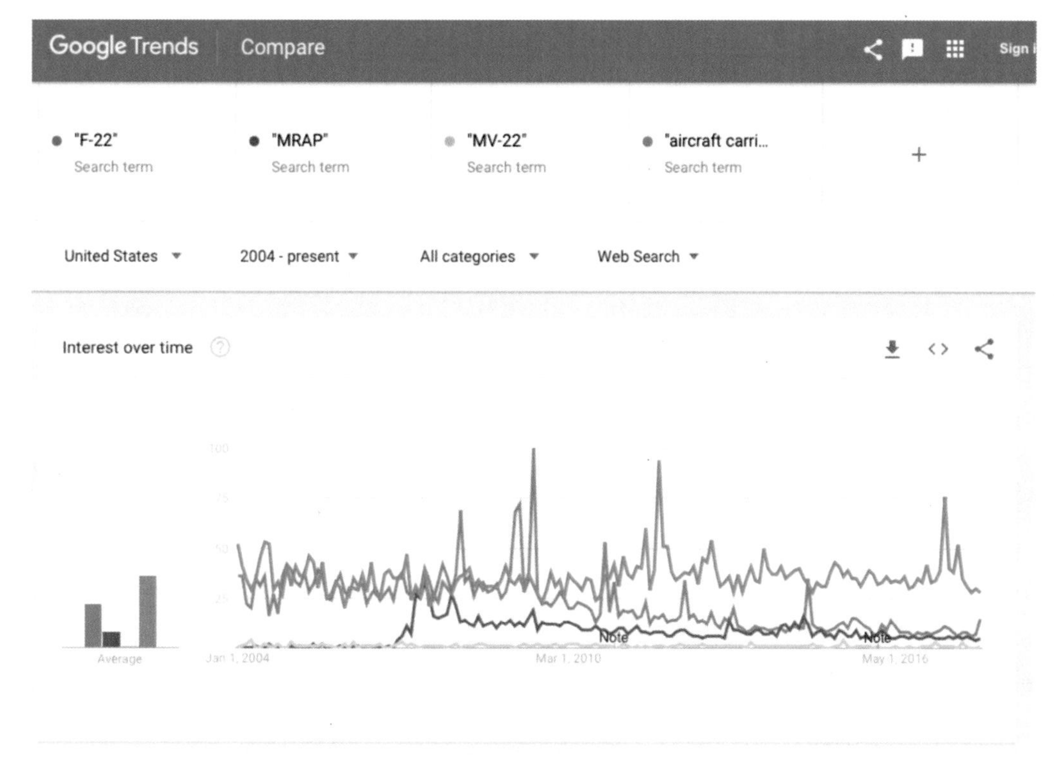

SOURCE: Google Trends search, December 19, 2017.

Figure 4.5. Relative Interest in F-22 Versus All Service Organizational-Level Presentations

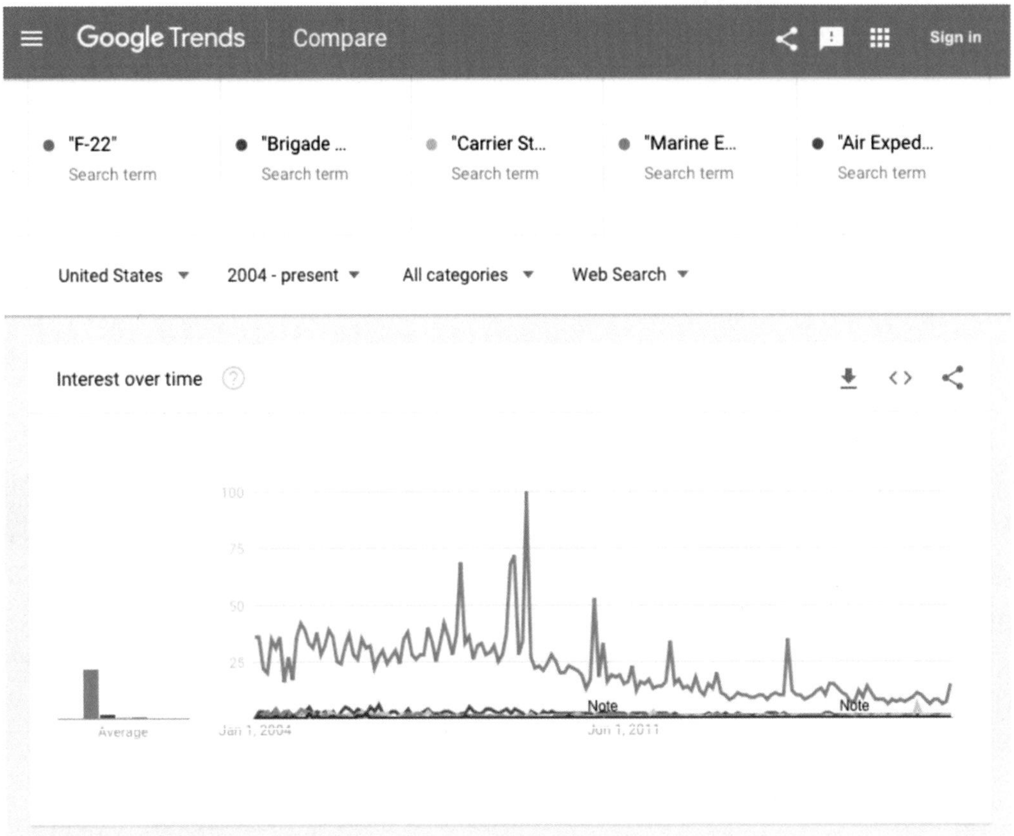

SOURCE: Google Trends search, December 19, 2017.

organizational-level constructs. As might be expected, public interest in the F-22 is vastly greater than organizational abstractions like the AEW, BCT, CSG, or MEU. This suggests that when USAF leaders speak or write about force presentation, they will more easily connect with the broader public by referencing iconic aircraft. References to the AEF and other organizational constructs are best left for more specialized audiences.

Two themes emerge from this analysis. First, force presentation is largely absent from USAF narrative documents and the writings of prominent airpower thinkers. Although this is understandable given the logic implicit in most airpower narratives, it makes for an incomplete and less compelling case for airpower. For that reason, force presentation should be integrated into future USAF narratives. Second, the public shows much greater interest in aircraft than in organizational structures and processes. Force presentation, therefore, should emphasize more accessible themes like the routine and rapid deployment of aircraft to trouble spots rather than the organizational mechanisms used for the deployments.

Chapter Five presents study conclusions and recommendations.

5. Conclusions and Recommendations

This chapter presents the main findings of the research and makes recommendations for USAF leaders.

Findings

Force Presentation Plays Distinct and Varied Roles in the Four Services

This paper defines *force presentation* as the preferred organizational construct through which a service offers and articulates its combat capabilities to combatant commanders, the president, and the nation. Although only the USAF uses this terminology, all of the services have force presentation constructs—and typically more than one, depending on the problem at hand. The present analysis has found that the services each use more than one construct to accomplish six key functions: (1) deploying forces, (2) employing forces, (3) sustaining operational effects, (4) sizing forces, (5) managing force rotations, and (6) articulating service purpose. For example, the Army uses the BCT construct to deploy and employ forces, the division and corps to sustain effects, the division to size the total force, and the ARFORGEN process to manage force rotations.

Force presentation is arguably most central to USMC and USN institutional identities and cultures because they articulate their service purpose primarily as providing forward presence and crisis response through rotational deployments of the MEU/ARG and CSG/ESG, respectively. In contrast, neither the Army nor USAF use force presentation constructs to articulate their service purpose.

The Squadron Has Been the Most Common USAF Force Presentation Construct for the Combat Air Forces

As was discussed in Chapter Three, a comparison of the squadron with the group, wing, or various AEF-related constructs over the history of the USAF finds the squadron to be the most frequently used across the six functions assessed in this report. The squadron has been the preferred deployment construct for all major conflicts, was paired with the group or wing for force employment, and was the most common force-sizing metric for the fighter and bomber force. The squadron appeared as the force presentation construct either alone or in tandem with the group or wing in 62 percent of the cases that were considered (see Table 3.2). Although the squadron is central to USAF culture and operations and possesses a surface familiarity to outside audiences, it probably is not the best vehicle for public outreach. As will be discussed below, for most people the squadron and other tactical echelons are abstractions that, by definition, have less visceral power than physical objects like aircraft, ships, and tanks.

Force Presentation Has Been Largely Ignored by Airpower Theorists and Historians

As has been documented in Chapter Four, force presentation concepts are largely absent from contemporary USAF planning documents and the writings of prominent airpower theorists. Force presentation is also rarely discussed in operational histories of the USAF at war. Although organization is of practical concern to all military organizations and leaders, the most prominent airpower theorists (Mitchell, de Seversky, Warden, and Deptula) have conceptualized and articulated the contribution of airpower using an effects-to-missions-to-targets-to-forces framework. This framework typically ends by highlighting specific airframes (e.g., the B-2), classes of aircraft (e.g., stealth bombers) and weapons types (e.g., the Joint Air-to-Surface Standoff Missile) among the required forces. This is helpful from a programmatic perspective, but is incomplete as a narrative because it leaves out both force presentation and basing (a critical enabler of global reach). Histories of the USAF at war also usually ignore organization except for descriptive purposes. Rather, they seek to tell a story that largely follows the effects-to-missions-to-targets-to-forces framework. They describe the air campaign objectives, planning process, and key participants; identify the critical targets; and tell the story of air combat, typically centered around particular individuals and platforms.

USAF Aircraft Are the Most Visible and Accessible Manifestation of Airpower

Although the modern USAF is an air, space, and cyber force, news accounts of USAF activities typically focus on the deployment or movement of aircraft. An August 14, 2016, *National Interest* article titled "B-1, B-2, and B-52 Bombers All Descend on Guam in a Massive Show of Force" is representative of news coverage of USAF deployments. Airmen, understandably, want to tell a broader story, but there is no getting around the visceral power of iconic aircraft. As was noted in Chapter Four, Google searches for aircraft like the F-22 greatly outnumber searches for the AEW, BCT, CSG, or MEU. Specialized audiences will care that the USAF presents forces as AES, AETF, or AEW, but the broader public has little use for such hard-to-visualize abstractions.

Recommendations

This research leads to three recommendations for USAF leaders:

- **Set realistic goals for any modifications to force presentation constructs.** This research shows that no single construct can address all six functions. Modifications should focus on better articulating USAF capabilities regarding particular functions such as "size forces" in a manner consistent with USAF history, culture, doctrine, and practices. Each service has developed force presentation concepts appropriate for its unique circumstances; none is easily transferable to other services. More broadly, a construct or set of constructs created by any one of the military branches is a relatively weak reed in the GFM system. Requests from CCDRs for forces will always have more weight in this process than will efforts by the services to constrain such demands. If USAF and other service leaders believe that CCDR demands for forces are unsustainable, they would be better served acting in concert to reform the GFM process so that it better highlights the readiness and other costs associated with current operational tempo.

- **Incorporate force presentation and basing into the USAF strategic narrative.** The most public American airpower narratives from the interwar years to today have followed a framework that begins with desired effects, derives missions and targets from those goals, and ends by identifying the forces needed to destroy those targets and accomplish the specified missions. This framework is logical but does not tell the whole story, leaving out both force presentation and basing. With regard to basing, the USAF is able to generate forces with such agility for a surprising reason: its global network of fixed facilities. From these locations the USAF integrates effects across domains via Air Operations Centers, conducts cybermissions, launches and controls space systems, and generates aircraft sorties. Thus, a more complete airpower narrative would use the following logic: desired effects to missions to targets to forces to force presentation to bases.
- **In public outreach, emphasize agility and scalability rather than organizational abstractions.** The best USAF force presentation public narrative is a simple one that involves no organizational references at all; the USAF is the most agile and scalable military force, able to respond rapidly and globally in packages as small as a single aircraft or as large as many hundreds of platforms supported by thousands of personnel. It does this through various organizational structures and processes, including the AEF and the widespread use of UTCs. But these "how" elements distract from the "what" and "why" narratives of widest interest and appeal. Agility and scalability are easily communicated concepts if connected to well-known physical objects like aircraft or satellites. Distinctive aircraft like the F-22 or B-2 are especially powerful icons and useful to illustrate agility in action. Organizational abstractions, no matter how important for internal processes, are no match for a clear, concise, and concrete narrative centered on USAF personnel deploying and employing advanced technologies on behalf of the nation.

Final Thoughts

This research suggests that the utility of force presentation as an overarching concept may be greatest for the two services whose strategic narratives are built on maintaining forward presence via rotational naval forces at sea. For the USMC and USN, their primary force presentation constructs—the MEU/ARG and the CSG, respectively—are easily tied to powerful images like the aircraft carrier at sea. In contrast, there is no single image that captures the essence of an AEW or other AEF-derived construct. For that reason, the USAF may need to take a more differentiated approach to force presentation, using the AEF model to deploy, employ, and sustain effects while at the same time employing more evocative and accessible themes in its public narrative.

Bibliography

Alexander, Robert M., "Force Structure for the Future," in Richard H. Shultz, Jr., and Robert L. Pfaltzgraff, Jr., eds., *The Future of Air Power in the Aftermath of the Gulf War*, Maxwell Air Force Base, Ala.: Air University Press, 1992, pp. 217–224.

Arnold, Henry H., "The Air Forces and Military Engineers," *The Military Engineer*, Vol. XXXIII, No. 194, December 1941, pp. 545–548.

Arnold, Henry H., and Ira C. Eaker, *Winged Warfare*, New York: Harper and Brothers, 1941.

Aspin, Les, *Report on the Bottom-Up Review*, Washington, D.C.: Office of the Secretary of Defense, October 1993.

Berger, Carl, ed., *The United States Air Force in Southeast Asia: 1961–1973*, Washington, D.C.: Office of Air Force History, 1977.

Bergerud, Eric M., *Fire in the Sky: The Air War in the South Pacific*, Boulder, Colo.: Westview Press, 2001.

Blechman, Barry M., and Stephen S. Kaplan, *Force Without War: U.S. Armed Forces as a Political Instrument*, Washington, D.C.: Brookings Institution Press, 1978.

Boyne, Walter J., *Beyond the Wild Blue: A History of the United States Air Force, 1947–2007*, 2nd ed., New York: St. Martin's Press, 2007.

Broughton, Jack, *Thud Ridge*, New York: J. B. Lippincott Company, 1969.

Cavas, Christopher P., "Top Marine: No Need to Change Deploying Groups," *Defense News*, August 10, 2016. As of February 6, 2018: https://www.defensenews.com/pentagon/2016/08/10/top-marine-no-need-to-change-deploying-groups/

Clodfelter, Mark, *The Limits of Air Power: The American Bombing of North Vietnam*, New York: The Free Press, 1989.

———, *Beneficial Bombing: The Progressive Foundations of American Air Power, 1917–1945*, Lincoln: University of Nebraska Press, 2010.

Coffey, Thomas M., *Decision over Schweinfurt: The U.S. 8th Air Force Battle for Daylight Bombing*, New York: David McKay Company, 1977.

Cook, Donald G., Robert Allardice, and Raymond D. Michael, Jr., "Strategic Implications of the Expeditionary Aerospace Force," *Aerospace Power Journal*, Vol. XIV, No. 4, Winter 2000. As of March 16, 2016: http://www.airpower.maxwell.af.mil/airchronicles/apj/apj00/win00/cook.htm

Craven, Wesley Frank, and James Lea Cate, *The Army Air Forces in World War II:* Vol. I, *Plans and Early Operations, January 1939 to August 1942*, Washington, D.C.: Office of Air Force History, 1983a.

———, *The Army Air Forces in World War II:* Vol. II, *Europe: Torch to Pointblank*, Washington, D.C.: Office of Air Force History, 1983b.

———, *The Army Air Forces in World War II:* Vol. V, *The Pacific: Matterhorn to Nagasaki*, Washington, D.C.: Office of Air Force History, 1983c.

———, *The Army Air Forces in World War II:* Vol. VI, *Men and Planes*, Washington, D.C.: Office of Air Force History, 1983d.

Crosby, Harry H., *A Wing and a Prayer: The "Bloody 100th" Bomb Group of the U.S. Eighth Air Force in Action over Europe in World War II*, New York: iUniverse.com, Inc., 2000.

Davis, Paul K., *Observations on the Rapid Deployment Joint Task Force: Origins, Direction, and Missions*, Santa Monica, Calif.: RAND Corporation, P-6751, June 1982. As of February 8, 2018: https://www.rand.org/pubs/papers/P6751.html

Davis, Richard G., *Immediate Reach, Immediate Power: The Air Expeditionary Force and American Power Projection in the Post Cold War Era*, Washington, D.C.: Air Force History and Museums Program, 1998.

———, *Anatomy of a Reform: The Expeditionary Aerospace Force*, Washington, D.C.: Air Force History and Museums Program, 2003.

Deaile, Melvin G., *The SAC Mentality: The Origins of Organizational Culture in Strategic Air Command, 1946–1962*, dissertation, Chapel Hill: University of North Carolina, 2007.

Deptula, David, "Air Force Transformation: Past, Present, and Future," *Aerospace Power Journal*, Vol. XV, No. 3, Fall 2001. As of March 16, 2016: http://www.airpower.maxwell.af.mil/airchronicles/apj/apj01/fal01/phifal01.html

———, *Effects-Based Operations: Change in the Nature of Warfare*, Arlington, Va.: Aerospace Education Foundation, 2001.

———, "Revisiting the Roles and Missions of the Armed Forces," statement before the Senate Armed Services Committee, November 5, 2015. As of February 8, 2018: https://www.armed-services.senate.gov/imo/media/doc/Deptula_11-05-15.pdf

de Seversky, Alexander P., *Victory Through Air Power*, New York: Simon and Schuster, 1942.

Douhet, Giulio, *The Command of the Air*, translated by Dino Ferrari, Washington, D.C.: Office of Air Force History, 1983. Originally published in Italian in 1921.

Finletter, Thomas K., *Survival in the Air Age: A Report*, Washington, D.C.: President's Air Policy Commission, 1947.

Foulois, Benjamin D., and C. V. Clines, *From the Wright Brothers to the Astronauts: The Memoirs of Major General Benjamin D. Foulois*, New York: McGraw-Hill, 1968.

Futrell, Robert F., *The United States Air Force in Southeast Asia: The Advisory Years to 1965*, Washington, D.C.: Office of Air Force History, 1981.

———, *The United States Air Force in Korea: 1950–1953*, Washington, D.C.: Office of Air Force History, 1983.

Futrell, Robert Frank, *Ideas, Concepts, Doctrine: Basic Thinking in the United States Air Force, 1907–1960*, Maxwell Air Force Base, Ala.: Air University Press, 1989.

Glosson, Buster, *War with Iraq: Critical Lessons*, Charlotte, N.C.: Glosson Family Foundation, 2003.

Goldfein, David, *CSAF Focus Area: The Beating Heart of the Air Force . . . Squadrons!* Washington, D.C.: Department of the Air Force, August 2016.

Google, Google Books, homepage. As of February 9, 2018: https://books.google.com

———, Google Trends, homepage. As of February 9, 2018: https://trends.google.com/trends/

Hallion, Richard P., *Storm over Iraq: Air Power and the Gulf War*, Washington, D.C.: Smithsonian Institution Press, 1992.

Haulman, Daniel, *Hitting Home: The Air Offensive Against Japan*, Washington, D.C.: Air Force History and Museums Program, 1999.

———, *Lineage and Honors History of the 1st Reconnaissance Squadron (ACC)*, Maxwell Air Force Base, Ala.: Air Force History Research Agency, November 2013.

———, *One Hundred Ten Years of Flight: USAF Chronology of Significant Air and Space Events 1903–2012*, updated ed., Maxwell Air Force Base, Ala.: Air Force History and Museums Program, 2015. Updated by Priscilla D. Jones and Robert D. Oliver.

———, *Table of Ten Oldest USAF Squadrons*, Maxwell Air Force Base, Ala.: Air Force Historical Research Agency, February 3, 2016.

Hennessy, Juliette A., *The United States Army Air Arm: April 1861 to April 1917*, Washington, D.C.: Office of Air Force History, 1985.

Hobson, Chris, *Vietnam Air Losses: United States Air Force, Navy and Marine Corps Fixed-Wing Aircraft Losses in Southeast Asia, 1961–1973*, Hinckley, England: Midland Publishing, 2001.

Hudson, James J., *Hostile Skies: A Combat History of the American Air Service in World War I*, Syracuse, N.Y.: Syracuse University Press, 1968.

Huntington, Samuel P., "National Policy and the Transoceanic Navy," *U.S. Naval Institute Proceedings*, Vol. LXXX, No. 5, May 1954. Last accessed on 2/28/18 at: https://blog.usni.org/posts/2009/03/09/from-our-archive-national-policy-and-the-transoceanic-navy-by-samuel-p-huntington

Hurley, Alfred F., *Billy Mitchell: Crusader for Air Power*, Bloomington: Indiana University Press, 1975.

Jaffe, Lorna S., *The Development of the Base Force 1989–1992*, Washington, D.C.: Office of the Chairman of the Joint Chiefs of Staff, Joint History Office, July 1993.

Johnson, David E., *Fast Tanks and Heavy Bombers: Innovation in the U.S. Army, 1917–1945*, Ithaca, N.Y.: Cornell University Press, 1998.

———, *Learning Large Lessons: The Evolving Roles of Ground Power and Air Power in the Post–Cold War Era*, Santa Monica, Calif.: RAND Corporation, MG-405, March 7, 2007. As of February 8, 2018: https://www.rand.org/pubs/monographs/MG405-1.html

Johnson, Herbert A., *Wingless Eagle: U.S. Army Aviation Through World War I*, Chapel Hill: University of North Carolina Press, 2001.

Joint Chiefs of Staff, *Department of Defense Dictionary of Military and Associated Terms*, Washington, D.C.: Joint Chiefs of Staff, JP 1-02, November 8, 2010. Amended through February 2016.

Joint Interagency Task Force South, homepage. As of June 17, 2016: http://www.jiatfs.southcom.mil

Krieger, Clifford R. "Fighting the Air War: A Wing Commander's Perspective," *Airpower Journal*, Vol. I, No. 1, Summer 1987. As of July 13, 2016: http://www.au.af.mil/au/afri/aspj/airchronicles/apj/apj87/sum87/krieger.html

Lambeth, Benjamin S., *The Transformation of American Air Power*, Ithaca, N.Y.: Cornell University Press, 2000.

———, *Air Power Against Terror: America's Conduct of Operation Enduring Freedom*, Santa Monica, Calif.: RAND Corporation, MG-166, 2005a. As of February 8, 2018: https://www.rand.org/pubs/monographs/MG166-1.html

———, *Carrier Air Power at the Dawn of a New Century*, Santa Monica, Calif.: RAND Corporation, MG-404, 2005b. As of February 8, 2018: https://www.rand.org/pubs/monographs/MG404.html

Larter, David B., "The U.S. Just Sent a Carrier Strike Group to Confront China," *Navy Times*, March 3, 2016. As of February 9, 2018: https://www.navytimes.com/news/your-navy/2016/03/03/the-u-s-just-sent-a-carrier-strike-group-to-confront-china/

Laslie, Brian D., *The Air Force Way of War: U.S. Tactics and Training After Vietnam*, Lexington: University Press of Kentucky, 2015.

Leighton, Richard M., *History of the Office of the Secretary of Defense:* Vol. III, *Strategy, Money, and the New Look, 1953–1956*, Washington, D.C.: Historical Office, Office of the Secretary of Defense, 2001.

Lendon, Brad, "U.S. Army Sending Armored Convoy 1,100 Miles Through Europe," *CNN Online*, March 14, 2015. As of February 8, 2018: https://www.cnn.com/2015/03/13/world/army-convoy-through-europe/index.html

Lewis, Adrian R., *The American Culture of War: The History of U.S. Military Force from World War II to Operation Iraqi Freedom*, New York: Routledge, 2007.

Looney, William R., "The Air Expeditionary Force: Taking the Air Force into the Twenty-First Century," *Airpower Journal*, Vol. X, No. 4, Winter 1996, pp. 5–9.

Mann, Edward C., III, *Thunder and Lightning: Desert Storm and the Airpower Debates*, Maxwell Air Force Base, Ala.: Air University Press, April 1994.

Maurer, Maurer, ed., *The U.S. Air Service in World War I:* Vol. I, *The Final Report and a Tactical History*, Washington, D.C.: Office of Air Force History, 1978a.

———, ed. *The U.S. Air Service in World War I:* Vol. II, *Early Concepts of Military Aviation*, Washington, D.C.: Office of Air Force History, 1978b.

———, ed. *The U.S. Air Service in World War I:* Vol. III, *The Battle of St. Mihiel*, Washington, D.C.: Office of Air Force History, 1978c.

———, ed., *Air Force Combat Units of World War II*, Washington, D.C.: Office of Air Force History, 1983.

———, *Aviation in the U.S. Army, 1919–1939*, Washington, D.C.: Office of Air Force History, 1987.

Meilinger, Phillip S., "Ten Propositions Regarding Air Power," *Airpower Journal*, Vol. X, No. 1, Spring 1996, pp. 1–18.

———, ed., *The Paths of Heaven: The Evolution of Airpower Theory*, Maxwell Air Force Base, Ala.: Air University Press, 1997.

Mellon, Jonathan, "Where and When Can We Use Google Trends to Measure Issue Salience?" *Political Science and Politics*, Vol. XLVI, No. 2, April 2013, pp. 280–290.

Miller, Roger G., *A Preliminary to War: The 1st Aero Squadron and the Mexican Punitive Expedition of 1916*, Washington, D.C.: Air Force History and Museums Program, 2003.

Mitchell, William, *Our Air Force: The Keystone of National Defense*, New York: E. P. Dutton and Company, 1921.

———, *Memoirs of World War I: From Start to Finish of Our Greatest War*, New York: Random House, 1960. Based on a 1926 manuscript.

———, *Winged Defense: The Development and Possibilities of Modern Air Power—Economic and Military*, Tuscaloosa: University of Alabama Press, 2009. Originally published 1925.

Momyer, William W., *Air Power in Three Wars*, Washington, D.C.: Office of Air Force History, 1985.

Mueller, Karl P., "Air Power," in Robert A. Denemark, ed., *The International Studies Encyclopedia:* Vol. I, Oxford: Wiley-Blackwell, 2010, pp. 47–65.

———, *Denying Flight: Strategic Options for Employing No-Fly Zones*, Santa Monica, Calif.: RAND Corporation, RR-423-AF, 2013. As of February 8, 2018: https://www.rand.org/pubs/research_reports/RR423.html

Nalty, Bernard C., *The Air Force Role in Five Crises: 1958–1965*, Washington, D.C.: U.S. Air Force Historical Division Liaison Office, 1968.

———, ed., *Winged Shield, Winged Sword: A History of the United States Air Force:* Vol. II, *1950–1997*, Washington, D.C.: Department of the Air Force, 1997.

Ochmanek, David, et al., *America's Security Deficit: Addressing the Imbalance Between Strategy and Resources in a Turbulent World*, Santa Monica, Calif.: RAND Corporation, 2015. As of February 8, 2018: https://www.rand.org/pubs/research_reports/RR1223.html

O'Connell, Aaron B., *Underdogs: The Making of the Modern Marine Corps*, Cambridge, Mass.: Harvard University Press, 2012.

Office of the Secretary of Defense, Historical Office, *Secretary of Defense Annual Reviews*, website. As of February 9, 2018: http://history.defense.gov/Historical-Sources/Secretary-of-Defense-Annual-Reports/

Olds, Robin, with Cristina Olds and Ed Rasimus, *Fighter Pilot: The Memoirs of Legendary Ace Robin Olds*, New York: St. Martin's Press, 2010.

Owen, Robert C., *Air Mobility: A Brief History of the American Experience*, Washington, D.C.: Potomac Books, 2013.

Oxford University Press, "Squadron," in *Oxford English Dictionary Online*. As of June 9, 2016: http://www.oed.com/view/Entry/188113?rskey=YUP9VD&result=1&isAdvanced=false#eid

Pape, Robert A., *Bombing to Win: Air Power and Coercion in War*, Ithaca, N.Y.: Cornell University Press, 1996.

Patrick, Mason M., *The United States in the Air*, Garden City, N.Y.: Doubleday, Doran and Company, 1928.

Pettyjohn, Stacie L., *U.S. Global Defense Posture, 1783–2011*, Santa Monica, Calif.: RAND Corporation, MG-1244-AF, 2012.

Polmar, Norman, ed., *Strategic Air Command: People, Aircraft, and Missiles*, Annapolis, Md.: Nautical and Aviation Publishing Company of America, 1979.

Ravenstein, Charles A., *Air Force Combat Wings: Lineage and Honors Histories, 1947–1977*, Washington, D.C.: Office of Air Force History, 1984.

Rickenbacker, Eddie, *Fighting the Flying Circus*, New York: Doubleday, 1965. Originally published 1919.

Royal Air Force, *1 (F) Squadron*, undated. As of March 31, 2018: https://www.raf.mod.uk/our-organisation/squadrons/1-f-squadron/

Rumsfeld, Donald H., *Annual Report to the President and the Congress*, Washington, D.C.: Office of the Secretary of Defense, 2003.

———, *Annual Report to the President and the Congress*, Washington, D.C.: Office of the Secretary of Defense, 2004.

Schehl, Matthew L. "2,300 California-Based Marines Just Deployed to the Middle East," *Marine Corps Times*, April 22, 2016. As of February 9, 2018: https://www.marinecorpstimes.com/news/your-marine-corps/2016/04/22/2300-california-based-marines-just-deployed-to-the-middle-east/

Schlight, John, *The War in South Vietnam: The Years of the Offensive, 1965–1968*, Air Force History and Museums Program, 1999.

Sherry, Michael S., *Preparing for the Next War: America Plans for Postwar Defense, 1941–45*, New Haven, Conn.: Yale University Press, 1977.

Shiner, John F., "Benjamin D. Foulois: In the Beginning," in John L. Frisbee, *Makers of the United States Air Force*, Washington, D.C.: Pergamon-Brassey's, 1989.

Smith, Jeffry F., *Commanding an Air Force Squadron in the Twenty-First Century: A Practical Guide of Tips and Techniques for Today's Squadron Commander*, Maxwell Air Force Base, Ala.: Air University Press, August 2003.

Smith, Patrick J., *Building the Eagle's Nest: Challenges in Basing the Air Expeditionary Force*, Maxwell Air Force Base, Ala.: School of Advanced Airpower Studies, 1997.

Smith, Perry McCoy, *The Air Force Plans for Peace: 1943–1945*, Baltimore, Md.: Johns Hopkins University Press, 1970.

Smith, Richard K., *Seventy-Five Years of Inflight Refueling: Highlights, 1923–1998*, Washington, D.C.: Air Force History and Museums Program, 1998.

Spencer, Otha C., *Flying the Hump: Memories of an Air War*, College Station: Texas A&M University Press, 1992.

Spires, David N., *Patton's Air Force: Forging a Legendary Air-Ground Team*, Washington, D.C.: Smithsonian Institution Press, 2002.

Stocking, Galen, and Katerina Eva Matsa, "Using Google Trends Data for Research? Here Are 6 Questions to Ask," *Pew Research Center*. As of December 19, 2017: https://medium.com/@pewresearch/using-google-trends-data-for-research-here-are-6-questions-to-ask-a7097f5fb526

Tate, James P., *The Army and Its Air Corps: Army Policy Toward Aviation, 1919–1941*, Maxwell Air Force Base, Ala.: Air University Press, 1998.

Thornhill, Paula G., *"Over Not Through": The Search for a Strong, Unified Culture for America's Airmen*, Santa Monica, Calif.: RAND Corporation, OP-386-AF, 2012. As of February 8, 2018: https://www.rand.org/pubs/occasional_papers/OP386.html

Toll, Ian, *Six Frigates: The Epic History of the Founding of the U.S. Navy*, New York: W. W. Norton and Company, 2008.

Tunner, William H., *Over the Hump*, Washington, D.C.: Air Force History and Museum Program, 1998.

U.S. Air Force, *Gulf War Air Power Survey:* Vol. I, *Planning and Command and Control*, Washington, D.C.: Department of the Air Force, 1993a.

———, *Gulf War Air Power Survey:* Vol. III, *Logistics and Support*, Washington, D.C.: Department of the Air Force, 1993b.

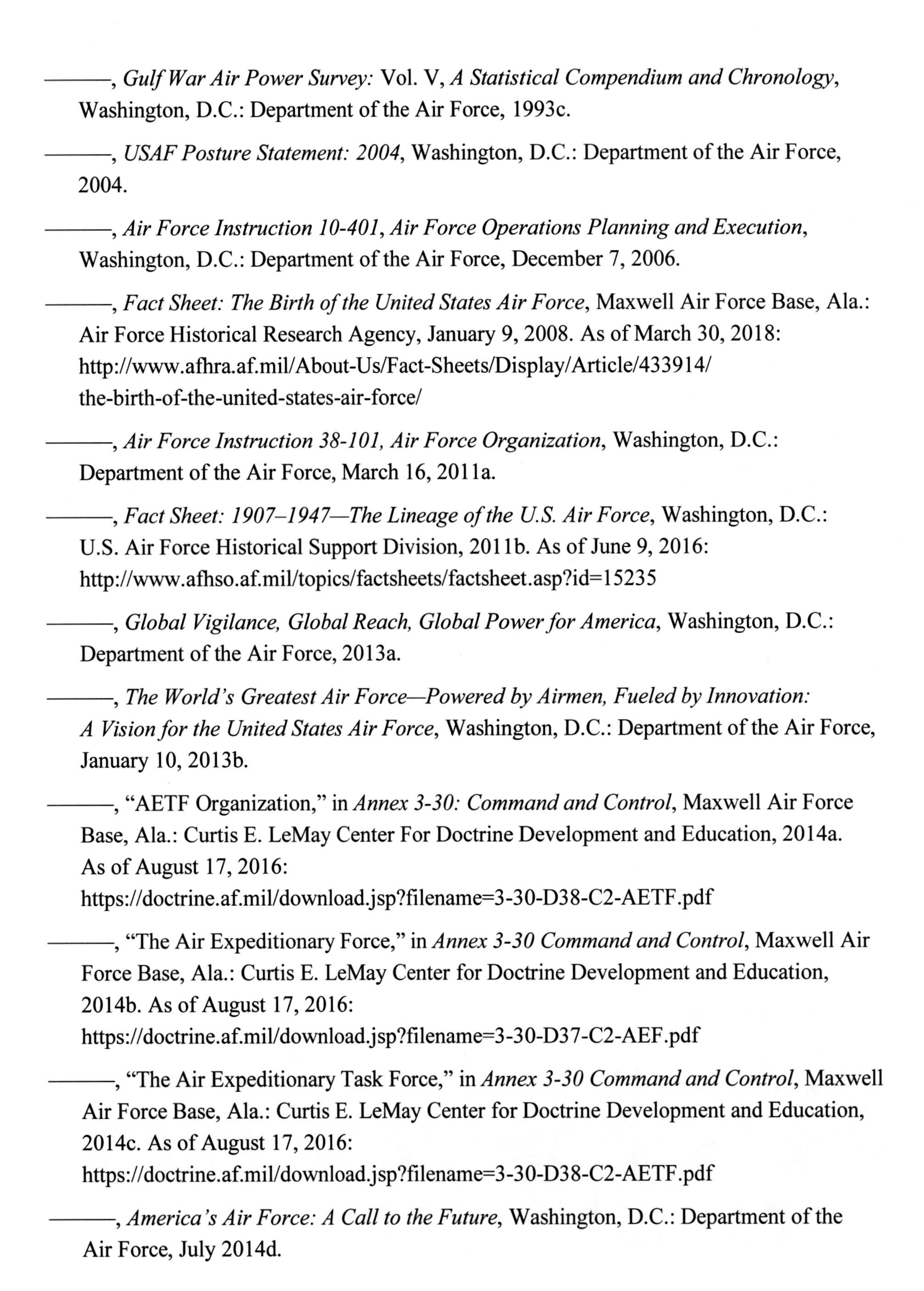

———, *Gulf War Air Power Survey:* Vol. V, *A Statistical Compendium and Chronology*, Washington, D.C.: Department of the Air Force, 1993c.

———, *USAF Posture Statement: 2004*, Washington, D.C.: Department of the Air Force, 2004.

———, *Air Force Instruction 10-401, Air Force Operations Planning and Execution*, Washington, D.C.: Department of the Air Force, December 7, 2006.

———, *Fact Sheet: The Birth of the United States Air Force*, Maxwell Air Force Base, Ala.: Air Force Historical Research Agency, January 9, 2008. As of March 30, 2018: http://www.afhra.af.mil/About-Us/Fact-Sheets/Display/Article/433914/the-birth-of-the-united-states-air-force/

———, *Air Force Instruction 38-101, Air Force Organization*, Washington, D.C.: Department of the Air Force, March 16, 2011a.

———, *Fact Sheet: 1907–1947—The Lineage of the U.S. Air Force*, Washington, D.C.: U.S. Air Force Historical Support Division, 2011b. As of June 9, 2016: http://www.afhso.af.mil/topics/factsheets/factsheet.asp?id=15235

———, *Global Vigilance, Global Reach, Global Power for America*, Washington, D.C.: Department of the Air Force, 2013a.

———, *The World's Greatest Air Force—Powered by Airmen, Fueled by Innovation: A Vision for the United States Air Force*, Washington, D.C.: Department of the Air Force, January 10, 2013b.

———, "AETF Organization," in *Annex 3-30: Command and Control*, Maxwell Air Force Base, Ala.: Curtis E. LeMay Center For Doctrine Development and Education, 2014a. As of August 17, 2016: https://doctrine.af.mil/download.jsp?filename=3-30-D38-C2-AETF.pdf

———, "The Air Expeditionary Force," in *Annex 3-30 Command and Control*, Maxwell Air Force Base, Ala.: Curtis E. LeMay Center for Doctrine Development and Education, 2014b. As of August 17, 2016: https://doctrine.af.mil/download.jsp?filename=3-30-D37-C2-AEF.pdf

———, "The Air Expeditionary Task Force," in *Annex 3-30 Command and Control*, Maxwell Air Force Base, Ala.: Curtis E. LeMay Center for Doctrine Development and Education, 2014c. As of August 17, 2016: https://doctrine.af.mil/download.jsp?filename=3-30-D38-C2-AETF.pdf

———, *America's Air Force: A Call to the Future*, Washington, D.C.: Department of the Air Force, July 2014d.

———, *Annex 3-30 Command and Control*, Maxwell Air Force Base, Ala.: Curtis E. LeMay Center for Doctrine Development and Education, 2014e. As of August 17, 2016: https://doctrine.af.mil/download.jsp?filename=3-30-D38-C2-AETF.pdf

———, *Strategic Master Plan Executive Summary*, Washington, D.C.: Department of the Air Force, 2014f.

———, *Strategic Master Plan: Strategic Posture Annex*, Washington, D.C.: Department of the Air Force, May 2015. As of March 31, 2016: http://www.af.mil/Portals/1/documents/Force%20Management/Strategic_Posture_Annex.pdf?timestamp=1434024340513

———, *A Guide to United States Air Force Lineage and Honors*, Maxwell Air Force Base, Ala.: Air Force Historical Research Agency, undated. As of March 31, 2018: http://www.afhra.af.mil/Portals/16/documents/Organizational-Records/AFD-090611-010.pdf

U.S. Department of Defense, *Quadrennial Defense Review 2001*, Washington, D.C.: Office of the Secretary of Defense, September 30, 2001.

———, *Quadrennial Defense Review 2014*, Washington, D.C.: Office of the Secretary of Defense, March 4, 2014.

Van De Walle, Curt A., *Back to the Future: Does History Support the Expeditionary Air Force Concept?* thesis, Maxwell Air Force Base, Ala., 2000.

Van Staaveren, Jacob, *Air Operations in the Taiwan Crisis of 1958*, Washington, D.C.: U.S. Air Force Historical Division Liaison Office, November 1962.

Van Staaveren, Jacob, Robert D. Little, and Wilhelmina Burch, *Air Operations 1958: Lebanon and Taiwan*, Newtown, Conn.: Defense Lion Publications, 2012.

Viccellio, Henry, Sr., "The Composite Air Strike Force 1958," *Air University Quarterly Review*, Vol. XI, No. 2, Summer 1959, pp. 2–17.

Vick, Alan J., *Proclaiming Airpower: Air Force Narratives and American Public Opinion from 1917 to 2014*, Santa Monica, Calif.: RAND Corporation, RR-1044-AF, 2015. As of February 8, 2018: https://www.rand.org/pubs/research_reports/RR1044.html

Ward, Clarissa, "U.S. Sends F-22 Warplanes to Romania," *CNN Online*, April 26, 2016. As of February 9, 2018: https://www.cnn.com/2016/04/25/europe/us-deploys-fighter-jets-to-romania/index.html

Warden, John, *The Air Campaign: Planning for Combat*, Washington, D.C.: Pergamon-Brassey's, 1989.

———, "The Enemy as a System," *Airpower Journal*, Vol. IX, No. 1, Spring 1995, pp. 40–55. As of December 19, 2017:
http://www.airuniversity.af.mil/Portals/10/ASPJ/journals/Volume-09_Issue-1-Se/1995_Vol9_No1.pdf

Warnock, A. Timothy, ed., *Short of War: Major USAF Contingency Operations, 1947–1997*, Maxwell Air Force Base, Ala.: Air Force History and Museums Program, 1983.

Weigley, Russell F., *The American Way of War: A History of United States Military Strategy and Policy*, Bloomington: Indiana University Press, 1977.

Weinberger, Caspar W., *Annual Report to the Congress, Fiscal Year 1985*, Washington, D.C.: Office of the Secretary of Defense, February 1, 1984.

Welsh, Mark, "Exit Interview," conducted by *Air Force Magazine* staff, June 17, 2016. Last accessed on 2/28/18 at:
http://www.airforcemag.com/DocumentFile/Documents/2016/WelshExitInterview.pdf

Werrell, Kenneth P., *Blankets of Fire: U.S. Bombers over Japan During World War II*, Washington, D.C.: Smithsonian Institution Press, 1996.

West, Michael B., *Evolution of the Marine Expeditionary Brigade*, Quantico, Va.: U.S. Marine Corps School of Advanced Warfighting, 1999.

White, Robert P., *Mason Patrick and the Fight for Air Service Independence*, Washington, D.C.: Smithsonian Institution Press, 2001.

Wilson, John B., *Maneuver and Firepower: The Evolution of Divisions and Separate Brigades*, Washington D.C.: U.S. Army Center of Military History, 1998.

Worley, D. Robert, *Shaping U.S. Military Forces: Revolution or Relevance in a Post–Cold War World*, Westport, Conn.: Praeger Security International, 2006.